The
Texas City
Disaster,
1947

The Texas City Disaster, 1947

Hugh W. Stephens

University of Texas Press
Austin

First edition, 1997

Requests for permission to reproduce material from this work
should be sent to Permissions, University of Texas Press,
Box 7819, Austin, TX 78713-7819.

∞ The paper used in this publication meets the
minimum requirements of American National Standard
for Information Sciences—Permanence of Paper for
Printed Library Materials, ANSI Z39.48-1984.

Frontispiece: Texas City, morning of 16 April 1947.
Moore Memorial Public Library, Texas City.

Library of Congress Cataloging-in-Publication Data

Stephens, Hugh W.

    The Texas City disaster, 1947 / Hugh W. Stephens. — 1st ed.

      p.   cm.

    Includes bibliographical references and index.

    ISBN 0-292-77722-1 (alk. paper). — ISBN 0-292-77723-x (pbk. : alk. paper)

    1. Fires—Texas—Texas City—History—20th century. 2. Disasters—Texas—
Texas City—History—20th century. 3. Texas City (Tex.)—History. 4. High Flyer
(Ship) 5. Grandcamp (Ship) 6. Wilson B. Keene (Ship) I. Title.

F394.T4S73 1997

976.4′139—dc20   96-35657

This book is

fondly dedicated to

the Emergency Program

Managers of Texas.

A day of wrath is that day,

a day of distress and anguish,

a day of ruin and devastation,

a day of darkness and gloom,

a day of clouds and thick darkness.

Zephaniah I:15

Contents

Illustrations

Preface

In terms of casualties, the Texas City disaster remains the worst industrial catastrophe in U.S. history. Its centerpiece is the explosion on 16 and 17 April 1947, sixteen hours apart, of ammonium nitrate fertilizer on two merchant ships. Because it occurred during the tremendous surge of chemical production and transportation that followed World War II, long before comprehensive precautions were embodied in the extensive governmental regulations that now encompass these operations, it constitutes the baseline hazardous materials disaster in the industrial culture of the United States. As such, it provides a potential benchmark for measuring progress in risk assessment and emergency management. Nonetheless, while issues of risk assessment and emergency management are necessarily present in this book, I have not attempted to extract lessons through the systematic application of these concepts, nor have I tested their applicability for analysis of subsequent disasters.

Replete with sudden death, mutilation, and property devastation, the disaster makes for disturbing but interesting history. During the immediate aftermath of the tremendous ship explosions, survivors exhibited a mixture of panic, heroism, and ingenuity in the face of imminent danger as well as confusion and ineptness. Surviving townspeople and residents of neighboring towns came to help, revealing grim determination to do whatever they could. The story concerns human nature in the workplace of the private sector and in government agencies at federal, state, and local levels. It demonstrates how tragedy can occur when employees, supervisors, managers, and bureaucrats charged with the safe handling of dangerous substances fail to fully consider the effect their actions might have upon others who share this responsibility. More than anything else, Texas City illustrates what can happen when complacency, neglect, ignorance, and even stupidity exist amid dangerous circumstances.

This disaster was a media event for radio and newspapers and riveted the attention of a shocked nation. It would have been particularly appropriate for the voyeurism of television, but this medium was then only in its infancy. After investigating the impact of television upon the federal government's response to Hurricane Andrew, two scholars concluded that it is a "voracious and insatiable medium" which finds disasters particularly appropriate for meeting its demands for drama and action.[1] One can only speculate about the impact the event would have had upon public opinion and government if television had been present to splash graphic images of the town's agony across the screens of America, accompanied by the provocative commentary of news reporters presenting their version of reality.

Curiously, the disaster and surrounding circumstances have never been investigated in their totality. Not even the technical reports submitted by fire and insurance investigators, the "experts" of the time, present information in a way that facilitates its incorporation into the overall context of the disaster. Apparently, many concluded that the sheer power of the explosions left little else to explain. Then again, given the complexity of the event, investigators may have been put off by the difficulty of making conclusive judgments. But the fact remains that hundreds died, thousands were injured, property worth millions of dollars was destroyed, and an entire community was severely disrupted by the ship explosions and their aftermath. Such results demand searching questions and forthright answers—if we cannot discern the truth about Texas City, we can neither identify those who are to blame nor recognize pitfalls to be avoided in the future.

How can we best identify and draw useful lessons from the pertinent conditions and actions that contributed to the disaster? For instance, how can we determine who was responsible for the safe transit of the fertilizer in terms of simple or culpable negligence? Can any of us recall an occasion when officials had the courage to step forward amid carnage and suffering and admit they were probably at fault? This certainly did not happen at Texas City. Even more intriguing is the fact that, except for damage suits entered in federal court by relatives of victims, the whole matter of culpability was quickly put aside. Consequently, our understanding of this horrible disaster rests on overly simple explanations and perhaps disinformation as well. Disinformation is not just a matter of lies;

it includes avoiding and not revealing awkward facts as well. A major premise of this book is that simple explanations of a complex event that happened in 1947 were just as misleading then as they are today. Was this deliberate or not? How much of the blame is attributable to a code of silence designed to discourage further inquiry? How much was the product of relative unfamiliarity with industrial disasters involving dangerous chemicals? Unfortunately, existing literature on the disaster does not provide much help in answering either question because it is largely confined to descriptions of events that triggered the ship explosions or to human interest stories about the aftermath.

Regardless of the presence or absence of a carefully drawn emergency plan, like every other disaster involving dangerous products, what happened at Texas City cannot be blamed on human error alone. This means that while "unsafe acts" on the part of officials require investigation, their significance must be assessed within the social and administrative context in which they occur. We need to learn whether or not information was disseminated about the potentially explosive quality of ammonium nitrate fertilizer, as well as who might or should have known this. We also need to understand how well safety standards were enforced at the docks.

The years since 1947 have witnessed improvements in emergency preparedness as well as in knowledge about hazardous products and safety programs. But the possibility of disaster is always relative to the seriousness of the threat. Whenever private sector operators and government officials are complacent, dilatory, or negligent about the dangers of storing and transporting hazardous substances and give only lip service to emergency preparedness, we are almost guaranteed to witness another Bhopal, *Exxon Valdez*, or even Texas City.

This book is designed to show what happened to people who were blind to the possibility of disaster and to provide explanations that may be helpful in understanding similar situations in our day and time. Chapter 1 provides a brief description of the explosions of the *Grandcamp* and *High Flyer*, the surrounding physical and cultural circumstances, and the state of preparedness for a major industrial emergency. The next four chapters describe the evolving crisis beginning on the morning of 16 April and ending on 22 April 1947. Chapter 2 takes up the chain of events leading to the explosion of the first ship, the *Grandcamp*. Chapter 3 details the effects of the blast and the reactions of survivors in Texas City and inhabitants of neighboring towns who rushed in to help. Chapter 4 describes events through the afternoon of the first day, focusing upon the difficulty both government and private sector officials encountered in formulating a coherent response amid the chaos. Chapter 5 describes the second day of the crisis, including the explosion of the *High Flyer* and subsequent events through the 22nd, when the crisis subsided and recovery began. Chapter 6 highlights continuing events over the next several months, including short-term recovery efforts and improvements in emergency preparedness inspired by the disaster. The final chapter provides a brief summary, including a discussion of the ramifications of poor preparedness for an industrial disaster, and assesses the question of negligence and culpability.

Like any study of an early disaster, this book cannot be complete and accurate in all respects. Texas City occurred before emergency management became an obligation of local government, before mandatory "after-action" reports, and before television gave us on-the-scene journalism. Moreover, nothing can wholly overcome the mind-numbing shock experienced by veterans of this event, the confused circumstances which followed the ship explosions, memories dimmed by the passage of half a century, and the death of key participants. But the story is told in considerable detail and, it is to be hoped, with insight that will enhance our understanding of this and ensuing disasters. It is offered to the reader in the same spirit as Thucydides offered his study of the Peloponnesian Wars: "It will be enough for me, however, if these words of mine are judged useful by those who want to understand clearly the events which happened in the past and which (human nature being what it is) will, at some point or other and in much the same ways, be repeated in the future."[2]

Acknowledgments

As the author of this book, I am responsible for its contents. But studies of this kind are never the product of one person alone. Many others contributed in important ways by providing information, evaluating the text, or offering moral support. The idea of writing this book came from an excellent paper on the disaster submitted by Gayle Praeger, a student in my emergency management seminar. I invited her to collaborate on further research, but, having moved on to greater things, she was unable to do so. A number of hardy survivors of the disaster consented to personal interviews, not only providing valuable insights but in some cases lending me their own memorabilia: Homer Allspach, W. T. (Andy) Anderson, Elsie and Charles Burd, Ken DeMaet, Helen Edrozo, Rosa Curry Eelbeck, Lynn Ellison, Fred Grissom, Mary Hunter, Dorothy Ware Jenkins, Rev. Frank Johnson, Warren B. Jones, James (Jimmy) Matthews, Kenneth T. Nunn, John Quinn, George Sanders, V. J. Schmitt, Charles Skipper, and John Weeks. Two veterans of the blast merit special mention for their generosity: Curtis Trahan, who was the mayor at the time of the disaster, and John Hill, who not only submitted to an extensive interview and provided valuable material from his own files but read an early version of the manuscript. These two gentlemen personify a description of Texas City's leadership offered by Agnes E. Meyer: "They have acquired the kindly, effortless self-confidence that characterizes all real leaders of men. Here was no vainglory, no boasting, no boosterism, but the generous, quiet atmosphere of strong personalities who do not need to prove their strength."[1] It is my fondest hope that the final product of my efforts not only accurately portrays the disaster but does honor to both survivors and victims of this holocaust.

A number of other individuals contributed in other ways. I am grateful to Gene Tullich for his help in locating retired Coast Guard personnel who were serving at the Galveston Station at the time of the disaster and to Sylvia Colvin for information on the Sabine-Neches Chiefs Association.

My son Andrew very kindly used his geographic information skills to generate the maps and a loading diagram of the *Grandcamp*. Charles Doyle, the present mayor of Texas City, merits thanks for his friendly support and encouragement. Casey Greene of the Rosenberg Library in Galveston and Colonel R. Scribner, USA (ret.), of the Texas State Guard Museum in Austin provided valuable material. I wish to thank George Grant for reading and commenting on parts of the manuscript, and I owe a deep debt of gratitude to Colonel Douglas G. Macnair, USA (ret.), who not only read the manuscript but made valuable suggestions about style and presentation. Thanks go to Dr. Glenn Aumann, head of the University of Houston Coastal Foundation, for providing travel funds for research.

Before I started this project, I already held three librarians in highest esteem—they are members of my family. But three more have been added. One is Barbara Rust, an archivist at the National Archives, Southwest Region, Fort Worth, Texas. With unfailing patience and efficiency, she dug out a voluminous amount of the material collected for *Dalehite v. United States*, the major damage suit filed in federal court in connection with the disaster. The other two are Susie Moncla, head librarian at the Moore Public Library of Texas City, and her colleague, Joanne Turner, the research librarian. These two ladies not only guided me through the considerable holdings on the disaster in their library but suggested the names of most of the informants listed above. Their friendship is one of the chief rewards of this effort. Finally, although I wish to publicly acknowledge the support and encouragement provided by my wife, Clare, only she and our children understand the importance of her unswerving loyalty through more than forty years of marriage.

The
Texas City
Disaster,
1947

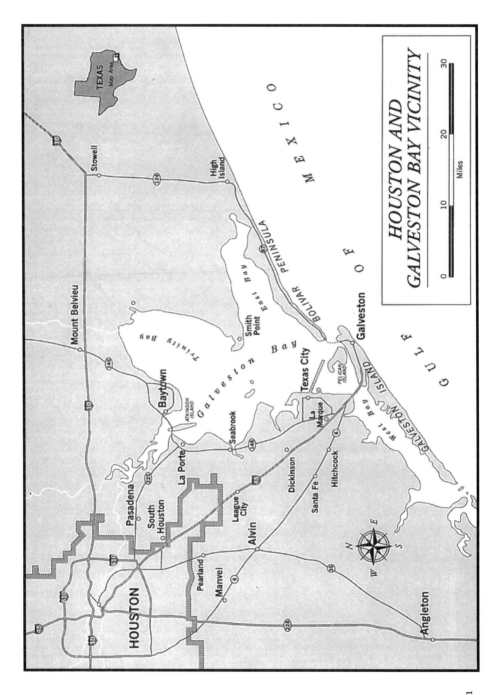

HOUSTON AND
GALVESTON BAY VICINITY

Miles

0   10   20   30

Map 1

1

The Blasts

The centerpiece of the Texas City disaster was the explosion
of ammonium nitrate fertilizer on board two Liberty ships
moored at the Texas City docks. What began as a small fire in
the hold of the *Grandcamp* quickly escalated into a cataclysmic
blast which disintegrated the ship, wreaking absolute havoc
within a radius of 2,000 feet. In the confusion that prevailed
throughout the remainder of the day, everyone overlooked the
potential danger of fertilizer in the other ship, the *High Flyer*.
The carnage was magnified when this ship exploded with equal
fury sixteen hours later. Much more than mistakes fighting the
fire on the *Grandcamp* was at issue. The event originated from
complacency about hazardous materials; the close physical proximity of docks, petrochemical facilities, and residences; and an
absence of preparation for a serious industrial emergency.

What Happened

The morning of 16 April 1947 dawned clear and crisp, cooled by a brisk
north wind. Just before 8:00 A.M., longshoremen removed the hatch covers on Hold 4 of the French Liberty ship *Grandcamp* as they prepared to
load the remainder of a consignment of ammonium nitrate fertilizer.
Some 2,300 tons were already on board, 880 of which were in the lower
part of Hold 4. The remainder of the ship's cargo consisted of large balls
of sisal twine, peanuts, drilling equipment, tobacco, cotton, and a few cases

of small arms ammunition. No special safety precautions were in force at the time.

Several longshoremen descended into the hold and waited for the first pallets holding the 100-pound packages to be hoisted from dockside. Soon thereafter, someone smelled smoke. A plume was observed rising between the cargo boards and the ship's hull, apparently about seven or eight layers of sacks down. Neither a gallon jug of drinking water nor the contents of two fire extinguishers supplied by crew members seemed to do much good. As the fire continued to grow, someone lowered a fire hose, but the water was not turned on. Since the area was fast filling with smoke, the longshoremen were ordered out of the hold.

While Leonard Boswell, the gang foreman, and Peter Suderman, superintendent of stevedores, discussed what action to take, the master, or captain, of the *Grandcamp* appeared and stated in intelligible English that he did not want to put out the fire with water because it would ruin the cargo. Instead, he elected to suppress the flames by having the hatches battened and covered with tarpaulins, the ventilators closed, and the steam system turned on. At the master's request, stevedores started removing cases of small arms ammunition from Hold 5 as a precautionary measure. As the fire grew, the increased heat forced the stevedores and some crew members to leave the ship. The *Grandcamp*'s whistle sounded an alarm that was quickly echoed by the siren of the Texas City Terminal Railway Company. Despite a strike by telephone workers, Suderman, seriously concerned by now, managed to reach the Fire Department and then called Galveston for a fire boat.

It was now about 8:30. At this point, growing pressure from the compressed steam fed into Hold 4 blew off the hatch covers, and a thick column of orange smoke billowed into the morning sky. Attracted by its unusual color and the sirens, several hundred onlookers began gathering a few hundred feet away at the head of the slip. Twenty-six men and the four trucks of the Volunteer Fire Department, followed by the Republic Oil Refining Company fire-fighting team, arrived on the scene and set up their hoses. A photograph taken at approximately 8:45 shows at least one stream playing on the deck of the *Grandcamp*, which was apparently hot enough to vaporize the water.

The *Grandcamp* on fire, about 8:45 A.M. National Archives, Southwest Region.

Around 9:00, flames erupted from the open hatch, with smoke variously described as "a pretty gold, yellow color" or as "orange smoke in the morning sunlight . . . beautiful to see."[1] Twelve minutes later, the *Grandcamp* disintegrated in a prodigious explosion heard as far as 150 miles distant. A huge mushroomlike cloud billowed more than 2,000 feet into the morning air, the shock wave knocking two light planes flying overhead out of the sky. A thick curtain of steel shards scythed through workers along the docks and a crowd of curious onlookers who had gathered at the head of the slip at which the ship was moored. Blast overpressure and heat disintegrated the bodies of the firefighters and ship's crew still on board. At the Monsanto plant, located across the slip, 145 of 450 shift workers perished. A fifteen-foot wave of water thrust from the slip by the force of the blast swept a large steel barge ashore and carried dead and injured persons back into the turning basin as it receded. Fragments of the *Grandcamp*, some weighing several tons, showered down throughout the port and town for several minutes, extending the range of casualties and property damage well into the business district, about a mile away. Falling shrapnel bombarded buildings and oil storage tanks at nearby refineries, ripping open pipes and tanks of flammable liquids and starting numerous fires. After the shrapnel, flaming balls of sisal and cotton from the ship's cargo fell out of the sky, adding to the growing conflagration.

The sheer power of the explosion and the towering cloud of black smoke billowing into the sky told everyone within twenty miles that something terrible had happened. People on the street in Galveston were thrown to the pavement, and glass store fronts shattered. Buildings swayed in Baytown fifteen miles to the north. The towering smoke column served as a grim beacon for motorists driving along the Houston-Galveston highway, some of whom immediately turned toward Texas City to help. In Texas City itself, stunned townspeople who started toward the docks soon encountered wounded persons staggering out of the swirling vortex of smoke and flame, most covered with a thick coat of black, oily water. Many agonizing hours were to pass before a semblance of order began to replace the shock and confusion caused by this totally unexpected and devastating event.

As the surge of injured quickly overwhelmed the town's three small medical clinics, the city auditorium was pressed into service as a makeshift first-aid center. Within an hour, doctors, nurses, and ambulances began arriving unsummoned from Galveston and nearby military bases. Serious casualties were taken to Galveston hospitals and later to military bases and even to Houston, fifty miles away. State troopers and law enforcement officers from nearby communities helped Texas City's seventeen-man police force maintain order and assisted in search and rescue.

The horror was not yet over. As help poured into Texas City, no one gave much thought to another Liberty ship tied up in the adjoining slip. The *High Flyer* was loaded with sulfur as well as a thousand tons of ammonium nitrate fertilizer. The force of the *Grandcamp*'s explosion had torn the *High Flyer* from its moorings and caused it to drift across the slip, where it lodged against another cargo vessel, the *Wilson B. Keene*. The *High Flyer* was severely damaged, but many of its crew members, although injured, remained on board for about an hour until the thick, oily smoke and sulfur fumes drifting across the waterfront forced the master to abandon ship. Much later in the afternoon, two men looking for casualties boarded the *High Flyer* and noticed flames coming from one of the holds. Although they reported this to someone at the waterfront, several more hours passed before anyone understood the significance of this situation, and not until 11:00 P.M. did tugs manned by volunteers arrive from Galveston to pull the burning ship away from the docks. Even though a boarding party cut the anchor chain, the tugs were unable to extract the ship from the slip.[2] By 1:00 A.M. on 17 April, flames were shooting out of the hold. The tugs retrieved the boarders, severed towlines, and moved quickly out of the slip. Ten minutes later, the *High Flyer* exploded in a blast witnesses thought even more powerful than that of the *Grandcamp*. Although casualties were light because rescue personnel had evacuated the dock area, the blast compounded already severe property damage. In what witnesses described as something resembling a fireworks display, incandescent chunks of steel which had been the ship arched high into the night sky and fell over a wide radius, starting numerous fires. Crude oil tanks burst into flames, and a chain reaction spread fires to other structures previously spared damage. When dawn arrived, large columns of thick, black smoke were visible thirty miles away. These clouds hovered over Texas City for several days until the fires gradually burned out or were extinguished by weary fire-fighting crews.

The *Grandcamp*'s explosion triggered the worst industrial disaster, resulting in the largest number of casualties, in American history. Such was the intensity of the blasts and the ensuing confusion that no one was able to establish precisely the number of dead and injured. Ultimately, the Red Cross and the Texas Department of Public Safety counted 405 identified and 63 unidentified dead. Another 100 persons were classified as "believed missing" because no trace of their remains was ever found.[3] Estimates of the injured are even less precise but appear to have been on the order of 3,500 persons.[4] Although not all casualties were residents of Texas City, the total was equivalent to a staggering 25 percent of the town's estimated population of 16,000. Aggregate property loss amounted to almost $100 million, or more than $700 million in today's monetary value. Even so, this figure may be too low, because this estimate does not include 1.5 million barrels of petroleum products consumed in flames, valued at approximately $500 million in 1947 terms.[5] Refinery infrastructure and pipelines, including about fifty oil storage tanks, incurred extensive damage or total destruction. The devastated Monsanto plant alone represented about $20 million of the total. Even though the port's break-bulk cargo-handling operations never resumed, Monsanto was rebuilt in little more than a year, and the petrochemical industry recovered quickly. One-third of the town's 1,519 houses were condemned, leaving 2,000 persons homeless and exacerbating an already-serious postwar housing shortage. Over the next six months, displaced victims returned as houses were repaired or replaced, and most of those who suffered severe trauma appear to have recovered relatively quickly. What could never be made good was the grief and bleak future confronting more than 800 grieving widows, children, and dependent parents.

Precursors

The fire that started in Hold 4 of the *Grandcamp* and the manner in which it was fought are typical of the way in which human error can initiate disasters. Mistakes occurred not only because safety at the docks was inadequate but also because officials were indifferent to the possibility of an industrial disaster and ignorant about the explosive potential of ammonium nitrate fertilizer. Still, accidents initiated by human errors—even those committed in ignorance—do not become disasters unless they interact with surrounding circumstances in a way that quickly magnifies the scale and severity of harm. The remainder of this chapter addresses the most significant of these circumstances, some of which existed elsewhere in the country as well as at Texas City. Often called precursors, these circumstances facilitated escalation of the disaster once the fertilizer in the *Grandcamp* caught fire.[6] One precursor was the presence of extensive petroleum refining and chemical production facilities together with large amounts of extremely flammable, explosive products stored at or near the waterfront. A second was the close proximity of docks and petrochemical facilities to each other as well as to part of the town's residential area. Still another was a mixture of cultural, psychological, and political attitudes held by responsible officials and the public; these helped sustain a general complacency about the possible dangers of a variety of chemical products passing through the port. Derived from this outlook was a fourth precursor—highly fragmented responsibility for safety at the waterfront and an absence of arrangements for coordinated response should a serious emergency occur. As the story unfolds, it will become apparent that this condition encouraged mutual ignorance about hazards among officials, hampered efforts to fight the fire on the *Grandcamp*, and became a major source of difficulty in coordinating response to the devastating effects of the ship explosions.

Physical Circumstances

Texas City is located on the west side of Galveston Bay, about ten miles north of Galveston. Port operations began in 1893, and ocean-going vessels started calling at the docks in 1904. Access remained a problem for about ten years until a channel thirty feet deep and three hundred feet wide was dredged and a protective dike constructed to prevent silting. The channel was later deepened to thirty-five feet to accommodate larger ocean-going vessels. The port derived some benefit from the fact that the route from the entrance to the Gulf of Mexico at Galveston Bay followed a straight-line course, and pilots were not required for ships calling there. By the close of World War I, port facilities consisted of three slips and two piers. After a serious fire in 1929, a large concrete two-story warehouse called Warehouse B was constructed, the grain elevator was modernized, and sprinkler systems were installed in most warehouses.

Land communications were also good. The Texas City Terminal Railway Company—hereafter called the Terminal Railway—was incorporated in 1921. It operated six miles of line connecting the docks to trunk routes of several major railways running between Houston and Galveston, which in turn provided access to markets and produce from the Midwest. As at many other ports at the time, there was no port authority; the Terminal Railway owned and operated storage and handling facilities located along a half-mile stretch of property on the west side of the turning basin. This meant the pattern of industrial development around the docks was set by the leasing policies of the Terminal Railway and the Mainland Company, the two principal landowners in the area. In addition to piers and warehouses for handling break-bulk commodities, several terminals for loading ocean-going tankers and coastal barges were constructed at the south end of the docks. Petroleum loading enjoyed the reputation of having good facilities and efficient service. Although break-bulk traffic provided a significant portion of the revenue, the port never had more than modest success as an outlet for midwestern agricultural products; consequently, transshipment of crude oil and other petroleum products constituted the greater share of the value of its traffic. When the Republic Oil Refining Company built a refinery in 1930, the town's second, Texas City began a rapid transition to an oil town. Four years later, Pan American Refining Corporation, a subsidiary of Standard of Indiana, built another refinery, this one capable of processing 25,000 barrels a day. During the mid-1930s

View of dock area before the explosions. National Archives.

other companies expanded their facilities, and more oil storage tanks appeared.

The preparation for war beginning in 1940 initiated an unprecedented expansion of industrial activity in this town of only 6,000 persons. That year, Carbide and Carbon Chemical Company constructed a large chemical plant for production of a wide variety of synthetic chemicals. Soon thereafter, a Seatrain intermodal terminal for transferring railway boxcars to ships was added, as well as a government-owned smelter to handle South American tin imports. The federal government's Defense Plant Corporation selected Texas City as the site of two of a total of eighteen alkylation plants and two of fourteen fluid catalytic cracking units built around the nation in anticipation of wartime demand for aviation fuel and lubricants.[7] In 1942, this agency also took over an old sugar refinery located across the North Slip from Dock O and rebuilt it for styrene production, a key ingredient in synthetic rubber. The plant was purchased by Monsanto in 1946 and continued to manufacture styrene and associated products. In 1947, the Terminal Railway installed facilities for handling bulk liquid petroleum products at the South Slip.

By 1947, then, Texas City had become an important port on the Gulf of Mexico, with extensive petrochemical production, storage, and shipment facilities. Sharing in a national surge of oil-based chemical production which rose from one to five million tons during the decade of the 1940s,[8] it was home to four major refineries, two aviation gasoline units, and two chemical companies. The refineries were capable of processing 150,000 barrels of oil daily. A cotton compress, grain elevators, and the Seatrain terminal were also part of dock operations. In 1946, the port cleared 3,907 ships and 13,441,000 net tons of cargo.[9] Although nonpetroleum traffic was declining, a variety of items continued to cross the docks, including sulfur, cotton, flour and grain, tin ore and blocks of refined tin from the smelter, as well as bars of copper from Arizona and slabs of zinc from Oklahoma. The town was described as "young and thriving."[10] Its population, swelled by migrants seeking work during a national recession, grew rapidly, reaching an estimated 16,000 persons by 1947. This figure could have been higher, but a housing shortage obliged many wage earners to live in nearby towns and commute to work.

The Texas City Disaster, 1947

Proximity

Economic growth created the second precursor of disaster. Increased prosperity and additions to the work force were welcome signs of progress in Texas City. No one seems to have considered the possibility that additional refining and storage facilities might raise safety problems by an order of magnitude. In view of the disaster, it is ironic that this growth increased the town's vulnerability, because development had long been guided by a form of zoning. Following a plan instituted by early developers from Duluth, Minnesota, and supported by Hugh B. Moore, the town's most prominent businessman and booster until his death in 1945, separate areas were designated for industrial, residential, institutional, and commercial development. A zoning ordinance enacted in January 1946 made this practice official.[11]

Unfortunately, zoning had negative consequences. The fact that refineries and tank farms were built immediately inland from the waterfront meant that these facilities and the tremendous quantities of hazardous materials were congregated in close proximity to each other as well as to the docks and the southern extremity of the town's residential area, where blacks and Hispanics lived. This dramatically increased the chance that a serious fire or explosion at or near the waterfront would initiate a chain reaction among facilities, with devastating effects on people and property. Although no consideration was given to its explosive quality, the potential for disaster increased even more when large quantities of ammonium nitrate fertilizer began to transit the port toward the close of 1945. Proximity meant the safe handling of fertilizer now affected not only those directly associated with the process—the Terminal Railway, stevedores, regulatory agencies, and masters of ships—but almost everyone and everything in the industrial area. The significance of physical proximity was not lost upon specialists who visited Texas City after the ship explosions. According to a staff report by the National Fire Protection Association, "The large loss of life occurred because of the immediate proximity of persons engaged in the industrial activity of the port and its exposed properties. . . . The huge property destruction occurred because of the direct exposure to blast damage of high-valued industrial plants and facilities."[12] Thus, a key ingredient for disaster was in place: people and property with no explicit relationship to fertilizer shipments were now unknowing hostages to fertilizer's explosive power.

Refineries and chemical companies adhered to standard safety practices in their manufacturing processes. For instance, storage tanks were equipped with suppressant systems and surrounded by containment dikes. Nevertheless, distances between facilities were, more frequently than not, minimal.

The Atlantic Pipeline Company, Humble Oil, Stone Oil, Republic Oil, Monsanto Chemical, Southport Petroleum, eleven warehouses, nine piers, a grain elevator, and approximately twenty-five blocks of residences were

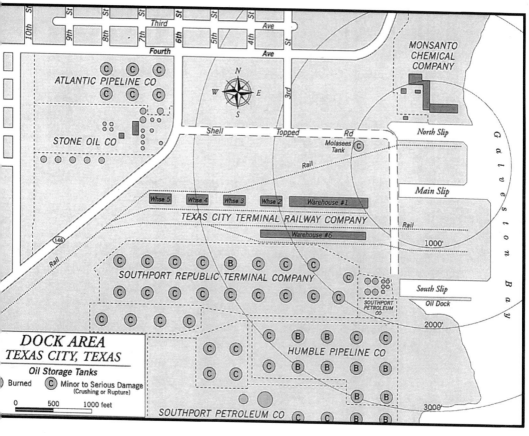

Map 2

all located within a one-mile radius of the North Slip (see Maps 2 and 4). Fire insurance specialists had estimated that not more than 20 percent of the industrial area of the terminal area and surrounding property was vulnerable to fire. But standard calculations and safety practices proved inadequate, because the ship blasts and resulting fires inflicted damage to about 90 percent of the area.[13] Visiting Texas City shortly after the explosions, Rear Admiral F. D. Higbee, then warden of the Port of Los Angeles, professed himself mystified as to why industry was concentrated so close to the waterfront when more land was available.[14]

Social Climate

Discussing why companies are often unable to cope with catastrophes, Ian Mitroff and Ralph Kilmann assert that the "culture" of organizations is a critical factor. In this context, culture denotes a mind-set, an orientation based on unwritten rules or tacit assumptions about the importance of seeking out threats and formulating tentative solutions in advance.[15] While the major refineries and companies at Texas City adhered to state-of-the-art safety practices within their fence lines, general complacency resulted in an indifference to the possibility of a wide-scale industrial disaster, leaving the town vulnerable to surprise and unprepared to deal with a catastrophe. Assessing the impact of this factor is a fascinating but elusive subject, because there is no explicit information about perceptions of danger entertained by private sector managers, municipal officials, or agencies such as the Coast Guard, much less the public. Seemingly, employees at the refineries and chemical plants understood that they were under some risk, but these concerns were offset by pay and benefits. Local newspaper stories prior to the disaster reveal no evidence of apprehension about any kind of hazardous material. Similarly, none of the survivors I interviewed remember anyone expressing worry about the dangers of hazardous materials. The fact that several hundred citizens gathered at the head of the North Slip and watched the *Grandcamp* burn is mute testimony to this truth as well as to general ignorance about the explosive potential of the fertilizer. Although there is some evidence that a few persons knew better than this, a brief report written several years after the disaster was essentially correct in stating that "people in Texas City were not conditioned to explosions; but they were conditioned to disregard very real threats to life and property."[16]

From today's perspective, it is difficult to understand why people living and working in close proximity to hazardous substances were indifferent to the possibility of catastrophe. Their "ethical perspective"[17] about the value of human life was certainly as good as anyone's, for the disaster produced an energetic, compassionate response by citizens and their neighbors, even to the extent of setting aside racial conventions in what was then a segregated southern town. But one must also understand that the United States immediately after World War II was a different world insofar as hazardous materials were concerned. Chemicals had made a vital contribution to Allied victory in the recent war and to the local economy as well. Scientists were confident about their knowledge, and the products of technology enjoyed virtually uncritical acceptance. Far in the future were spectacular and disturbing events such as the near catastrophe at Three Mile Island, Bhopal's horrendous death toll, and the massive oil spill from the tanker *Exxon Valdez*. It was well into the 1960s before a diffuse but very real sense of dread about the risks of new technology coalesced around an emerging counterculture and a broad-based environmental movement.[18] And only recently have the dangers of technological events begun to equate with natural ones in the public mind, even in communities where large quantities of hazardous chemicals are present. Not until 3,000 persons were killed at Bhopal and the 1986 Superfund Amendment and Reauthorization Act was passed did petrochemical companies expand their scope of awareness and planning to include adjacent communities. Even now, some question the total candor and realism of their actions.

Before the accident, Texas City's citizens felt secure because such a disaster had never happened and they could therefore assume it never would. From time to time, explosions and fires had occurred at refineries and tank farms, and fires had broken out on ships, but none had been really serious. The absence of any response plan for a major emergency by refineries, the Terminal Railway, and/or the municipal government suggests an important fact: nobody, neither private sector companies nor municipal agencies, was seriously concerned about an industrial disaster. In the words of one postmortem: "the people of Texas City had given little thought to industrial disasters nor had they taken their organization for natural disasters very seriously."[19]

It is likely that Texas City's unprecedented prosperity also contributed to prevailing indifference about the possibility of an industrial disaster. At a time when most of the rest of the nation was mired in recession, Texas City was a boomtown. Judging by later experience, communities where the social and political environment emphasize growth at the expense of safety or social justice are especially susceptible to catastrophe.[20] There is no evidence that a measurable part of the citizenry dissented from this perspective; although labor unions were active, matters concerning operational safety do not seem to have marred relations with management. Moreover, Texas City was basically a one-industry town where almost a third of the labor force was employed at refineries, oil terminals, or chemical plants.

Safety and Emergency Preparedness

The final precursor is the state of safety and emergency preparedness in and around the waterfront. Both were grossly deficient, considering the enormity of the dangers. Their inadequacy demonstrates that officials responsible for the safe transit of hazardous materials through the port were not sufficiently aware of possible hazards, and they had not instituted measures that would have reduced vulnerability to fire and explosion. Exactly who was culpable in this respect and to what extent is in fact a major concern of this study. Neglect had several ramifications, all of them bad. Not only were accidents more likely, but so was surprise. Without preparations, little chance existed that anyone could cope with the effects of an accident quickly enough to prevent it from escalating into a disaster, or that once begun, response would be effective enough to avoid its worst effects. Visiting Texas City soon after the disaster, a Los Angeles fire chief perceptively noted that "while there was no lack of succor for the victims . . . the existence of a Disaster Plan would have speeded the organization of several agencies involved."[21]

In a way, Texas City was not unacquainted with disaster. Early in its history the town received much of the force of the great hurricanes of 1900 and 1915 as they roared out of the Gulf of Mexico and into Galveston. In 1943, a smaller hurricane left considerable damage in its wake. This annual threat had prompted the Galveston chapter of the American Red Cross to draw up countywide relief plans that included Texas City. Nonetheless, a planning meeting of the town's own chapter in March had failed to attract much interest even among those who attended.[22] In any case, hurricane plans were of limited value in alerting citizens to the threat of industrial emergencies and creating support for measures which would have been appropriate for the ship explosions. Many people perceive hurricanes as acts of God or forces of nature and therefore discount antecedent economic and social conditions, so important in influencing the outcome of industrial disasters. Moreover, while not preventable, hurricanes usually provide some advance warning, allowing those in the projected path to flee or take protective measures. As the chairman of the local Red Cross chapter later observed: "We weren't prepared for anything like this. In a hurricane, you have some warning and can get your organization ready. But we had no warning, and our organization was scattered all over town."[23]

His statement hits upon an important truth—industrial disasters are different from natural disasters. Even when good safety measures are in place, it is impossible to predict exactly when explosions or fires will trigger an event. They happen with little or no warning and inflict most of their damage at the outset. This means that the quality of response is basically conditioned by whatever preparations are already in place, including contingency plans and designated resources. Because large amounts of highly flammable and explosive petrochemical products were stored in close proximity to each other and the waterfront, safety standards should already have been high. When ammonium nitrate fertilizer shipments began arriving in late 1945, these standards should have been reevaluated; however, they were not.

Safety and preparedness for industrial emergencies were also hindered by the fragmented nature of waterfront operations. Fragmentation derives from prevailing differences in skills, language, activities, and even per-spective between maritime and landside environments. Such differences create an organizational "fault line" at the water's edge which obscures a common perception of hazards among the variety of organizations and hinders cooperation. Texas City was no different from other ports in this respect, but it lacked a port authority that encompassed both environ-ments and had the capability to act quickly and effectively in an emer-gency. Terminal Railway officials were later to testify that company policy was to take no responsibility for safety aboard ships, even when they moored at the docks. As a result, nothing existed beyond personal friend-ships to facilitate cooperation between the Terminal Railway, the munic-ipality, and neighboring refineries, allowing the situation to get quickly out of hand after the fertilizer on the *Grandcamp* caught fire. Many friend-ships disappeared in the holocaust which followed.

Not all the responsibility for poor safety practices belonged to the lo-cals. Standards for handling hazardous materials at Texas City and other ports might have been better if governmental supervision over loading hazardous materials had been present. The lack of such supervision was a critical shortcoming. During World War II, the federal government had hurried "a staggering array of hazardous materials, products, and pro-cesses" into production, including styrene for synthetic rubber and high-octane aviation gasoline, both produced at Texas City, and ammonium nitrate as well.[24] Where risks of shipping and storing such materials were known to be high, military authorities successfully instituted and super-vised strict standards for safety and security. The record was excellent: in fact, 50 million tons of explosives were conveyed by rail during World War II without any loss of life directly attributable to transportation prob-lems.[25] But little thought was given to transportation safety when military supervision was curtailed at the war's end and transport was transferred into commercial channels. Several months after the disaster, the Coast Guard's Merchant Marine Council concluded that this lapse in supervi-sion was a contributing factor to the accident. "This lack of familiarity with the regulations," stated the council, "may be due to the fact that the regulations became effective just a few months before the war and that during the war, the loading of ships was in the main supervised by military personnel."[26] While this statement is accurate up to a point, it is critically incomplete. What the council neglected to mention was that the Coast Guard itself, in charge of administering Admiralty law, was responsible for filling the gap in supervision but had not done so.

Much has been learned about industrial disasters since 1947. Along with an expanding body of knowledge about the dangerous qualities of chemical compounds and new technology, the safety of hazardous materials production and transportation has been enhanced by means of comprehensive regulations. In the past ten years or so, private sector associations and government agencies have raised safety standards and added elaborate requirements with respect to contingency planning, response, and recovery. But disasters involving hazardous chemicals still occur, and often do so with devastating consequences. While this account may provide some perspective on other disasters, its focus is upon what happened at Texas City. The examination is oriented around several key questions never even raised by the U.S. Coast Guard board inquiring into the Texas City disaster and barely mentioned, if at all, by fire and insurance investigators. First, how did an accident in the form of a small fire in a ship escalate into an explosion which killed or injured the equivalent of one-quarter of the town's population and devastated a substantial portion of its industrial and residential property? Second, why did the explosion surprise so many of those on the scene? Third, were the explosions of the *Grandcamp* and *High Flyer* just a matter of "bad luck," as so many believed at the time, or, as has already been implied, did conditions at the docks leave the area ripe for serious trouble? Should not at least some officials responsible for transporting fertilizer have known what could happen? Fourth, why was the threat of the fertilizer on the *High Flyer* overlooked for almost twelve hours after the *Grandcamp* blew up? Fifth, what were the effects of a virtual absence of preparations for a large-scale industrial disaster at the docks and in the municipality?

2

The *Grandcamp*

OVERVIEW

This chapter examines conditions and actions which precipitated the explosion of the *Grandcamp*. The first part discusses three conditions that contributed to the high potential for disaster at Texas City: the apparent ignorance and indifference of officials about the hazardous potential of the fertilizer, the absence of governmental supervision over dock operations, and the weaknesses of safety management and practices there. The second part of the chapter demonstrates how these conditions allowed a small fire which began in a hold of the *Grandcamp* to quickly escalate into a major disaster.

Sins of Omission

There was no particular reason why the arrival of the *Grandcamp* on 11 April for a consignment of fertilizer was more likely to bring on disaster than visits of other ships before it. Indeed, the potential for disaster had existed for fifteen or sixteen months. For one thing, a variety of officials at manufacturing plants, the Terminal Railway, the U.S. Bureau of Mines, the Interstate Commerce Commission, and the Coast Guard, as well as shipping agents and ship captains, who should have known about the fertilizer's potential danger, failed to take interest, and as a result no warning was ever issued. For another, no regulatory body at any level of government, federal, state, or municipal, exerted its authority to supervise the safe handling of fertilizer or any other dangerous substance as these moved across the docks. By default, safety was left to the Terminal Railway, and its standards proved critically deficient. Such endemic neglect poses awkward questions. Since ammonium nitrate, which had about half the explosive power of TNT, was a major component of military munitions, why was its potential danger in fertilizer ignored by so many different agencies? To what extent was the basic duty of government to exercise its police powers and ensure the safety of the public abrogated? How carefully was the fertilizer handled in its transit through the port? What was the condition of safety at the docks, and what arrangements were in place to respond to a serious emergency? Small matter that prior to 16 April, thousands of tons of ammonium nitrate fertilizer had crossed the docks and been loaded into ships without trouble; the answers to these questions show how a small shipboard fire quickly escalated into an explosion that killed or injured thousands of people and caused extensive damage to their property.

The 7,176-ton *Grandcamp* was similar to hundreds of other Liberty-class cargo ships built by the United States during World War II. Recently handed over to the French line, it had a French crew temporarily commanded by Captain Charles de Guillebon. Already on board was a varied cargo of compressed cotton, sisal binder twine, tobacco, shelled peanuts, oil well and agricultural machinery, and sixteen cases of small arms ammunition. The loading of a 2,500-ton consignment of fertilizer at Dock O began the afternoon of 14 April. By 7:00 P.M. on the 15th, Hold 2 was full, but rain caused an interruption, leaving 600 tons still to be stowed in Hold 4. The loading was to be finished during the morning of the 16th. While this was in process, the *Grandcamp*'s turbine casing was removed for inspection of the blades, rendering the *Propeller* inoperative but allowing the ship to function normally otherwise. A similar ship belonging to Lykes Brothers Steamship Lines, the *High Flyer*, was moored in an adjoining slip. After taking on 961 tons of fertilizer on the 14th, it had been moved from Pier O to Pier A to receive disassembled boxcars. In one of its holds were 1,050 tons of sulfur previously loaded at Galveston.

The Texas City Disaster, 1947

Ammonium nitrate fertilizer was a relatively recent addition to U.S. commodity exports. A crystalline substance, whitish in its original color but usually tinged with brown because a small amount of clay was added to prevent caking, it was manufactured at several army ordnance plants in the Midwest which had formerly produced ammonium nitrate for munitions. It had the advantage of being relatively inexpensive and was richer in nitrogen than other combinations such as ammonium sulfate and sodium nitrate. Approximately 1.4 million tons had been produced in 1941, about half of which were exported. By 1947, annual production had reached almost 1 million tons, compared to 18,000 tons ten years previously. Postwar shipments to Western Europe were part of an effort by the United States to enhance food production and speed recovery from the war.[1] The fertilizer was shipped by rail to several ports, including Galveston, New Orleans, and Baltimore, as well as Texas City. Since this traffic had begun in late 1945, approximately 75,000 tons had crossed the Texas City docks. The particular consignment which exploded at Texas City was produced at the Cornhusker, Iowa, and Nebraska ordnance plants under contract to the Emergency Export Corporation, an agency of the federal government. These plants operated under the supervision of the Army's Bureau of Ordnance. As was the usual practice, ownership of this consignment was transferred from the original shipping agent to another firm buying on behalf of the French Shipping Council. J. B. Latta was the forwarding agent for the fertilizer loaded aboard the *Grandcamp*, destined for Brest, France.

The manner in which the fertilizer made its way through the port was similar to the way in which most other bulk commodities were handled. Railway boxcars containing the 100-pound, six-ply paper sacks were shunted onto a siding running along the land side of Warehouse O. Laborers removed the bags from the boxcars and took them into the warehouse by lift truck for temporary storage prior to transfer onto a vessel. The final phase involved placing the bags on wooden pallets, called "drafts," in layers eight high. Lift trucks conveyed drafts to the docks, where they were then transferred into the holds of ships by hoist or winch.

Apparently, no one at the docks was aware that the fertilizer posed any danger. The port's insurance underwriters placed no restrictions on the

Fertilizer stowed on a Liberty ship. Moore Memorial Public Library, Texas City.

product, and the Terminal Railway issued no special handling instructions. Laborers and stevedores testified after the disaster that the fertilizer was considered to be in the same class as cement or flour. Apparently, the Terminal Railway had not inquired about how to handle the fertilizer when shipments began, but W. H. "Swede" Sandberg, vice-president of the company, later asserted that after he noticed that the fertilizer came from an army ordnance plant he had asked a representative of one of the plants if the material was explosive and was told it was not.[2] This was a critical misstatement by the company representative—ammonium nitrate, the major ingredient, is an oxidizing agent and can catch fire or explode under certain circumstances. Ignorance was sustained by incomplete and ambiguous information about exactly what these conditions were. Having investigated past mishaps, Matthew Braidech, director of research for the National Board of Fire Underwriters, reported that before the Texas City disaster, prevailing scientific opinion held that the fertilizer was inert and would not catch fire or explode under "ordinary conditions."[3] Although a 1941 Army ordnance safety manual listed ammonium nitrate as a high explosive, according to a publication of the U.S. Department of Agriculture, "Commercial fertilizer mixtures containing ammonium nitrate require no special precautions regarding explosions."[4] There was a caveat, however: this publication explicitly states that ammonium nitrate can be exploded by a strong initial impulse, such as detonation of TNT, but this is difficult to accomplish *except in a confined space that retains heat and pressure*. An official of the Emergency Export Corporation had reported much the same thing just a few months prior to the disaster: "ammonium nitrate is not considered explosive under transportation conditions or when . . . stored in paper bags, by itself and apart from other explosive substances."[5] Inspectors from the U.S. Bureau of Mines who rushed to Texas City after the *Grandcamp* exploded found "a complete lack of understanding of the hazardous nature of ammonium nitrate fertilizer in the presence of fire and open flames was practically universal, even including experts manufacturing or handling this material."[6]

The substance could catch fire much more readily than it could explode. Tests made during World War II indicated that foreign matter, such as the small amount of clay and the petroleum, rosin, and paraffin mixture added to prevent caking or the coated paper used for packaging, lowered the temperature at which ammonium nitrate would begin to burn. The possibility of combustion at lower temperature when in contact with other substances may explain why the *High Flyer* blew up, since the ship already held a large quantity of sulfur. Subjected to heat, ammonium nitrate is fairly stable up to its fusion point of 338 degrees Fahrenheit, when it begins to melt, emitting combustible and toxic gas. Between this level and 390 degrees, as thermal breakdown reaction begins, oxidization becomes self-generating. At higher temperatures gas production increases, and at about 575 degrees, ammonium nitrate begins to burn in puffs of flame, giving off smoke of a light brown or orange-copper color. When heated quickly to a high temperature, as would occur in a confined space, the reaction to continued breakdown shifts from decomposition to detonation, causing what one specialist termed a "molecular dissociation and liberation of high amounts of energy in an instant"—to most of us, an explosion.[7] High-velocity detonators such as TNT that temporarily produce high temperatures will accomplish the same result. The difficulty of causing an explosion except in an enclosed space, where heat and pressure can build quickly, is illustrated by the fact that fertilizer bags in Warehouse O adjacent to the *Grandcamp* did not catch fire when the ship exploded. Even before Texas City, it was known that the best way to extinguish burning ammonium nitrate was to reduce the temperature by applying copious amounts of water.

By 1947, then, experience and scientific investigation had established that ammonium nitrate could catch fire or explode under certain circumstances. Dr. R. E. O. Davis, a chemist with the U.S. Department of Agriculture, testified that from 1943 onward, ammonium nitrate was understood to be a high explosive.[8] But few others knew what these circumstances were. Consider the criticism of officials from the U.S. Bureau of Mines, whose immediate dispatch to Texas City constitutes an implicit admission that their agency had a special understanding of the danger. If everyone at Texas City was ignorant of its explosive potential, then the Bureau of Mines bears some responsibility for this situation since it issued licenses for explosives and had the power to regulate distribution of the fertilizer. Subsequent investigation revealed that ammonium nitrate fertilizer had been instrumental in eleven serious fires or explosions since 1916. The worst had occurred at Oppau, Germany, in 1921, when an extra-powerful charge used to blast loose caked fertilizer caused a huge explosion, wiping out the town, killing 580 persons, and injuring 2,000. In 1943, Underwriters Laboratories had informed the War Production Board that tests demonstrated that ammonium nitrate coated with organic matter was more sensitive to detonation than the pure product and would be even more so if the substance was hot. Underwriters' recommendation for additional testing was ignored.[9] The next year, several persons were killed in an explosion at the Wolf Creek Army ordnance plant in Tennessee when lubricating grease and oil fell into molten ammonium nitrate. Although Army Ordnance instructed the investigating officer to prepare an article on this danger, this was not done, nor was the report on the accident publicly disclosed. A news story published in Texas papers on 4 May 1947 asserted that the Port of Houston had prohibited the loading of fertilizer in August previous to the disaster when chemical analysis revealed the danger and the port decided it could not comply with its underwriters' requirement that a special fire wall be built for storage. One survivor of the Texas City disaster reported that the chief of Texas City's Volunteer Fire Department may have been aware of the combustible properties of nitrate, but apparently this concern was never communicated to Terminal Railway officials in a way that caused improvements in safety practices. The results were cogently described by one specialist:

> The Texas catastrophe should be an unforgettable lesson for this industrial age . . . in which a considerable number of processes and resulting products . . . have become accomplished facts before a full opportunity was presented to fully develop proper safeguarding measures and suitable controls. The grim urgencies of our last war have crowded into the past five years at least two decades of normal research and development.[10]

Bureaucratic failure to identify the hazardous potential of products and circulate this information to users was a significant reason why "proper safeguarding measures and suitable controls" were absent at Texas City. Although the Coast Guard had designated the fertilizer as a "dangerous substance" in 1941, it was not classified as an explosive and not listed as such in the table of common hazardous chemicals prepared by the National Fire Protection Association. Any perception of danger was probably diminished because the bags in which it was packed did not carry the yellow warning label mandated by the Bureau of Explosives and the American Association of Railroads. Instead, printed in black were the words "Fertilizer," "Ammonium Nitrate," "Nitrogen 32.5 percent," together with information about weight and size. The board of investigation convened by the Coast Guard immediately after the disaster concluded that the manner in which fertilizer was shipped from army ordnance plants violated a section of Interstate Commerce Commission regulations on the transportation of explosives and other dangerous articles because the fertilizer was described under an unauthorized shipping name. But because it was not designated as an "explosive," shippers were not required to provide written notice to the masters of vessels which received it as cargo. A variety of other practices did not have to be followed either, including the presence of trained hatch, fire, and cargo guards, fire and safety equipment, or standby tugs during loading.

As fundamental as it was, the lack of information about the hazardous potential of the fertilizer was only one source of danger. Safety was further compromised by another void—the absence of supervision over the movement of hazardous substances through U.S. ports. Because the state of Texas did not assert jurisdiction, a regulatory void over hazardous materials operations at ports appeared when the Coast Guard did not take over supervision after the army withdrew its control of munitions transfers at the end of World War II.

The supine posture of the Coast Guard about this matter is difficult to understand. If this agency considered the fertilizer a "dangerous substance" in April 1941, then it was still dangerous six years later. No one attached any significance to the fact that during the war, ammonium nitrate exports were restricted to two ports where loading was under Army control. The Coast Guard did not regularly inspect operations involving any sort of dangerous cargo and asserted that it would do so only when violations were brought to its attention. In his deposition for the Dalehite case, the federal court suit which consolidated damage claims against the United States filed by relatives of victims of the ship explosions, the chief of the Office of Merchant Marine Safety admitted that the Coast Guard lacked funds for inspection of dangerous cargo and relied upon owners and operators to comply with regulations.[11] The flaw in this posture is illustrated by the fact that the chief of the Coast Guard's Merchant Vessel Inspection Division testified that shipping agents were not paying attention to dangerous cargo regulations at the time. The report of the Coast Guard's own investigating board states that the rules about explosives and dangerous cargoes were "self-regulating and policing."[12] Testifying under oath, the Coast Guard captain of the port at Galveston admitted that his organization was responsible for enforcing regulations concerning explosive and dangerous articles loaded on ships but that he had been unaware that ammonium nitrate fertilizer was being shipped through Texas City.[13] Although the language is tentative, the judgment of an investigator of the disaster suggests that an absence of supervision over safety matters was a critical shortcoming: "Perhaps the disaster is chargeable to the relaxation of strict rules and rigid inspections around our shipping centers as an aftermath letdown of the war. For with the passing [termination] of security regulations and routine inspections, a hazard was actually created under the conditions of volume handling of a dangerous cargo."[14] In effect, amid widespread indifference about the potential dangers of fertilizer, safety depended on the very fragile assumption that circumstances would always be "ordinary."

The only other governmental entity with authority over the dock area

Firefighters at *Grandcamp*, about 8:45 A.M. National Archives.

at Texas City was the municipality. Swede Sandberg later testified that the municipality and the Terminal Railway had no specific understanding concerning the former's jurisdiction at the docks even though the greater part of the area lay within the corporate limits of the city.[15] If the municipality had exercised law and order functions throughout Terminal Railway property at the time, given the absence of a port authority, there was still no entity capable of ordering a burning ship into the turning basin or attending to numerous other marine safety considerations. Although the president and vice-president of the Terminal Railway had traditionally occupied influential positions in community economic development and politics, no evidence exists that these officers ever intervened to shield the company's property from regulation. Law enforcement at the docks, including suppression of drunkenness and disorderly conduct, devolved onto guards employed by the Terminal Railway. The jurisdictional autonomy of the docks from municipal authority is illustrated by the fact that when stevedores struck the Terminal Railway the year before the disaster, law enforcement officers from other communities, not the Texas City police, were employed to maintain order.

All sorts of omissions may have contributed to the explosion of the *Grandcamp*. One specialist who investigated the disaster characterized safety at the docks as "complete laxity all along the line."[16] Without federal, state, or municipal supervision and lacking a port authority, such matters devolved upon the shipping agents, the masters of ships, and the Terminal Railway. For instance, shipping agents were responsible for ensuring that masters whom they represented posted fire watches when necessary, and masters had the obligation to refuse dangerous cargo which arrived in broken containers. In his deposition to the FBI, Henry Boone, head of fire prevention for the Los Angeles Fire Department, stated, "It was a known fact that the Texas City Terminal Railway Company was lax, and this knowledge encouraged ship owners to use its facilities since, as a general rule, they object to safety regulations."[17] Although water mains, hydrants, and pump boxes installed at the docks did meet fire insurance requirements, the Terminal Railway had no safety engineering specialist or officer with specific fire suppression duties, such matters being left to department supervisors. In what must surely be one of history's most prescient observations about a hazardous situation, one James Gavin told a meeting of the National Maritime Union in New York City just two days before the disaster that he didn't like the looks of a Gulf port named Texas City because inadequate safety precautions made the harbor a "natural" for a major explosion disaster.[18]

Inadequate safety measures created numerous opportunities for the fire and subsequent explosion. Among these were fertilizer spilled from the bags (the loose product is much more flammable than the packaged product) as well as sacks of fertilizer stored in contact with flour, grain, and other materials. Some consignments arrived still hot from the manufacturing process, resulting in bags that were brittle and that sometimes split or broke apart as they were transferred from boxcars to the warehouse or from the latter onto ships. In early 1946, Terminal Railway officials complained to ordnance plant officials about the condition of the sacks, apparently to good effect; a representative who visited Texas City a month before the disaster returned to tell his superiors that workers had told him that bags were no more than warm and few were broken. In the interests of safety, spilled contents were supposed to be swept up and put back in sacks, and the sacks were then resewn. This practice appears to have been followed reasonably well in the transfer from railway cars to the warehouse alongside Dock O, but testimony of stevedores suggests that no one performed this task on the docks or in the holds of ships as these were loaded. Torn bags and those ripped open in the process of being hoisted into the holds were allowed to remain in that condition, with the result that their contents often spilled onto the dunnage at the bottom of the hold.

Cigarette smoking was another in the long line of management failures, of particular significance because it posed the greatest threat of fire. Although red-lettered signs prohibiting smoking were posted on the docks, in warehouses, and on ships, the FBI concluded that the practice was nevertheless widespread. As was the case with spilled fertilizer, enforcement appears to have been better around the warehouse, where Terminal Railway officials were in charge, than on the docks and ships, where union loyalty and the authority of the masters were important. Peter Suderman, outside superintendent for Suderman Stevedores, stated that union policy mandated that workmen caught smoking on a ship were taken off the job, but he could not remember that this had ever happened.[19] The City Commission, Texas City's governing body, had passed an ordinance banning smoking on the docks in 1922. William Ladish, who had served as police chief for ten years by 1947, remembered that John Fuerst, chief of the Terminal Railway guard force killed in the *Grandcamp*'s explosion, was worried about violations of this ordinance but had never filed a complaint. Watchmen in the employ of the shipowner were supposed to enforce the no smoking rule, but no one on the *Grandcamp* was performing this duty on the morning of 16 April. Longshoremen working in the holds would occasionally lay lighted cigarettes down on the sacks of fertilizer. Two days before the disaster, a fire among fertilizer sacks in another of the *Grandcamp*'s holds which lasted for fifteen minutes was apparently started by a burning cigarette.

Lack of supervision of the cleanliness of the *Grandcamp*'s cargo area is another example of "laxity all along the line." The evidence on this point is not conclusive, but the presence of broken sacks and loose fertilizer would have allowed the fire in Hold 4 to start much more easily. The cargo underwriter is responsible for the cleanliness of holds prior to loading. With respect to the *Grandcamp*, this duty belonged to Captain Emmett Dierlan, who was in charge of inspecting ships in the Galveston area for the underwriters. Captain Dierlan told the Coast Guard board of investigation that he had inspected the *Grandcamp* at about 7:00 P.M. on 13 April and found Holds 2 and 4 clean. The trouble with this assertion is that these holds were already partly loaded by then, and he was unable to produce the required certificate. Some stevedores later told FBI agents that these holds were indeed clean before loading, that the dunnage—layers of boards laid at right angles to each other—was good and that heavy paper had been placed over this to ensure that fertilizer bags did not come into contact with the skin of the ship. But another man who helped lay the dunnage recalled that paper and scraps of board had been left on the bottom of the hold. It is also possible that some trash could have accumulated during loading, since it was common practice for stevedores to stuff the remains of torn bags between the cargo boards and the skin of the ship after these broke.[20]

The same indifference obviated preparations for any sort of serious emergency. In the view of Admiral Higbee, unregulated private ownership and operation of port facilities not only were a major source of safety failures but were responsible for the absence of contingency planning for an emergency.[21] That is, in a situation where facilities and residences were crowded together and large amounts of highly flammable or explosive chemical products were present, effective response demanded an organization whose scope was sufficiently broad to identify mutual vulnerabilities and that was capable of coordinating the application of municipal, petrochemical industry, and Terminal Railway resources. While Admiral Higbee's judgments were made with the benefit of hindsight, the train of events which began at 8:00 on the morning of 16 April confirms the importance of shortcomings in safety and preparations for an emergency.

Sins of Commission

The day crew of stevedores who were to finish loading the fertilizer on the *Grandcamp* went to work at 8:00 A.M. An eight-man gang assigned to work Hold 4 first removed the hatch covers put in place the evening before to protect the cargo from rain spawned by an approaching cold front. The four men on the offshore side of the hold began stowing bags left the day before, while the other team waited for the first draft from the docks. Surviving members of the gang asserted that no one noticed anything unusual when they entered the hold, but about ten minutes later someone smelled smoke. Julio Luna and Bill Thompson located a blue plume rising near the skin of the ship on the inshore side. Moving several bags, they discovered a small fire among other packages about ten or fifteen feet below. Luna recalled later that one end of a bag of fertilizer appeared to be glowing. At this point, the gang called for water. The contents of two jugs of drinking water had no effect, and the fire seemed to spread. Several members of the ship's crew, which was entirely French, descended into the hold with two soda acid fire extinguishers, but application of their contents seemed only to make the fire burn faster. A hose was lowered into the hold but not turned on.

As the fire grew and smoke thickened, gang foreman Leonard Boswell ordered the stevedores onto the deck. James Fagg, the walking foreman, and Peter Suderman rushed to the scene. All the survivors testified that a ship's officer, probably the master, Captain Guillebon, appeared at this point. He stated in reasonably intelligible English that he did not want to put out the fire with water because it would ruin the cargo; the ship's steam smothering system would be used instead. Made in total ignorance of the fertilizer's explosive potential, this was the critical decision that initiated thermal decomposition of the fertilizer and triggered the explosion. Significantly, no one else knew enough to warn the master of the possible consequences. As the ship's whistle screamed out an alarm of three short blasts and one long, the hatches were battened down and ventilators closed. Almost immediately after the steam system was activated by the chief engineer of the *Grandcamp*, the hatch cover on Hold 4 began to jump up and down as though considerable pressure were pushing it from below.

At the master's request, several longshoremen were ordered into Hold 5 to remove some cases of small arms ammunition, but this task was discontinued because of the extreme heat coming from the fire in Hold 4. About 8:30, as the Terminal Railway siren sounded, the stevedores were ordered off the ship. Walking or running, the less fearful went into Frank's Café at the end of the pier while others drove to the union hall in town to await further instructions. One stevedore is supposed to have jokingly told the gathering crowd, "If that fertilizer blows up, be sure and call us."[22] Probably more concerned that the heat would set off the ammunition, most of the *Grandcamp*'s crew began leaving the ship.

Even at this early stage the absence of standing emergency procedures resulted in confusion and more mistakes that allowed the crisis to escalate further. After leaving the *Grandcamp*, Peter Suderman made two telephone calls, both completed with considerable difficulty because local telephone operators were on strike. One was directed to the fire chief, requesting assistance; although the warehouses had sprinklers, and yard hydrants served by underground fire mains were available, the Terminal Railway lacked its own fire-fighting apparatus. The Fire Department's log shows that this telephone message was received at 8:37, so at least thirty minutes had passed from the time the fire was discovered to the time an alarm was turned in. Suderman's other call was placed to the Binnings

Company in Galveston, requesting a fire boat, since neither a fire boat nor tugs were immediately available to pull the ship into the turning basin, where it would be less of a threat. There was no question of moving the ship under its own power since repairs were under way in the engine room. Someone else, possibly Grant Wheaton of the Terminal Railway, called William Ludlow of the G & H Towing Company in Galveston, stating that a ship was on fire and requesting two tugs.[23] Approximately fifty minutes were required for the journey, but the two tugs assigned to the task, *Albatross* and *Propeller*, were unable to leave immediately since they were in the final stages of docking a ship. Galveston's fire boat, *The City of Galveston*, was also summoned but left after *Albatross* and *Propeller*, for it had not reached the channel leading into the turning basin when the *Grandcamp* exploded.

The first two of the Texas City Volunteer Fire Department's four trucks and twenty-seven men of its force appeared at the docks about 8:45. Shortly thereafter, the other two trucks arrived. Thermal decomposition of burning ammonium nitrate was apparently now well advanced. Foamite teams from Monsanto Chemical and Republic Oil were also present to provide assistance. As hoses were stretched across the dock from the hydrants, the hatch cover on Hold 4 blew several feet into the air. Smoke variously described as reddish brown or orange billowed out, together with pieces of burning paper. A picture taken by the chief engineer of the *High Flyer* shortly before 9:00 shows that the deck was hot enough to vaporize water from the hoses.

As the fire on the *Grandcamp* continued, it began to attract attention, some prudent and some foolish. Across the North Slip at the large Monsanto Chemical Company plant, 451 personnel on the morning shift were already at work. Robert Morris, the assistant plant manager, noticed the commotion on the *Grandcamp* and brought the company fire-fighting crews to standby status. Watching the fire burn from an office window, Tom Womack casually remarked to Fred Grissom, a fellow engineer, "That ship's got ammonium nitrate on it and is going to blow up."

On the *High Flyer*, moored at Pier A in the Main Slip, hatches had been removed to finish loading the disassembled boxcars. The *Grandcamp*'s fire prompted the master, Moseley Petermann, to sound the ship's general fire alarm and muster the crew to their fire stations. Hatches were closed and covered. Petermann assumed that the Fire Department was extinguishing the blaze, so his main concern was the swirling sparks and dense smoke which occasionally enveloped the *High Flyer*. He also requested the walking foreman of the stevedores working on his vessel to call the Lykes Brothers Steamship Lines office in Galveston and inform them that the ship in the next pier was on fire and precautionary measures had been taken.[24]

H. B. Williams, safety engineer for Pan American Refining Corporation, later told how he was on the way to Frank's Café, located near the head of the North Slip, when he noticed the orange-colored smoke. Thinking that a chlorine drum on the ship might have ruptured and seeing the smoke drifting southward over the company docks, he quickly went there and instructed all personnel to leave the area immediately.[25] When he detected the odor of nitric acid, he and another company official got in a car, intending to go and get gas masks for the volunteer firemen. "The rising column of orange smoke in the morning sunlight was beautiful to see," he remembered. "The run of the fire trucks directed attention to the spectacle, and within minutes all roads leading to the docks were congested with spectators bent on securing a closer view. . . . Even as we left the area, with siren screaming and red light flashing, it was necessary to nose people out of the way for perhaps the first hundred yards from the slip."[26] Apparently no one at the scene even contemplated the possibility that the fertilizer might explode. Reportedly, the only persons concerned about an explosion were crewmen and stevedores who knew about the small arms ammunition on the *Grandcamp*. Even jokes have some substance to them, and if a crewman really expected the fertilizer to explode and actually made the statement to the crowd mentioned above, then for whatever reason some sensed danger. Certainly, Terminal Railway officials did not expect an explosion. The casual attitude toward possible danger prevailing at the time was revealed in a court deposition given by Swede Sandberg: "Well, I merely turned around in my office chair and looked out the office window and I could see this orange smoke ascending skyward and thought not too much about it, we had ships on fire before that were always controlled within a short time, and I felt no difference in this instance."[27] Homer Allspach, who had assumed duties as a production superintendent at Carbide and Carbon Chemical Company only six weeks before, was more cautious. Even though he was two miles from the docks, having some knowledge of nitric acid and not liking the looks of the smoke, he decided to stay away.

The fire, particularly the unusual color of the smoke, attracted people who had no reason to be anywhere near Pier O, especially after Galveston radio station KGBC broadcast a message asking everyone to stay away from the docks. Efforts by police to keep the curious out of the area were feeble and ineffective. Sensing danger, John Fuerst tried to get workers lining the side of the warehouse to leave but was unsuccessful as well.

Notified of the fire shortly after 8:30, police chief William Ladish dispatched officer W. A. Reeves to the dock area and directed J. M. Bell and J. M. Wright to set up roadblocks at nearby street intersections, the latter just behind the Monsanto plant. But even spectators driving automobiles parked and filtered through. By 9:00, several hundred people were concentrated in the immediate vicinity of Dock O. One group along Pier O consisted primarily of Terminal Railway employees and stevedores. A second was directly across the slip from the *Grandcamp* in front of the Monsanto Chemical Company plant. The curious who had no business at the waterfront comprised the third element and were gathered in an open area at the head of the slip, approximately 1,000 feet from the ship. Perhaps as many as five hundred spectators were easily within sight of the ship, some school-aged children, since crowded Texas City schools were operating on a half-day schedule. Several hundred other less-venturesome persons gathered along the road beyond the police roadblocks, still within 2,000 feet of Pier O. Two light airplanes appeared overhead.

Shortly after 9:00, the fire seemed to abate slightly. A few people drifted back to their jobs. The tugs summoned from Galveston were almost in view of the docks. But all the while, steam was being fed into the confines of Hold 4, increasing heat and pressure and raising the temperature of the fertilizer toward the critical point. By this time, most of the crew had left the *Grandcamp* and were standing on the pier while Captain Guillebon took muster. Swede Sandberg had just put down the telephone after telling Jim Tompkins of Lykes Brothers Steamship Lines in Galveston that the burning ship posed no danger to his company's ship, the *High Flyer*. At

9:12, Tom Womack's casual prediction came true—the *Grandcamp* disintegrated in a tremendous explosion heard over a hundred miles away and powerful enough to register on a seismograph in Denver, Colorado. There are lots of careless people in the world but even more innocent bystanders; something that had heretofore concerned only the Terminal Railway company, the shipping agent, the stevedores, and the master of the *Grandcamp* suddenly and tragically became the business of the entire town.

No one knows with certainty exactly what triggered the explosion. The most plausible hypothesis involves fuel oil. Liberty ships of this type had a large vertical tank for Bunker C fuel oil located between Holds 3 and 4, as well as bottom tanks along the keel. It is possible that the extreme heat of the burning fertilizer in the hold eventually caused the bulkhead between the two compartments to give way, dumping a large amount of fuel oil on the molten fertilizer and setting off the explosion.[28] This theory is seemingly confirmed by the aforementioned explosion in early 1944 at the army's Wolf Creek ordnance plant in Tennessee when lubricating grease and oil fell into molten ammonium nitrate, killing several workers. It should be noted, however, that fertilizer explosions have occurred when oil was not involved, including during tests conducted at the Army's Picatinny Arsenal after the Texas City disaster.[29]

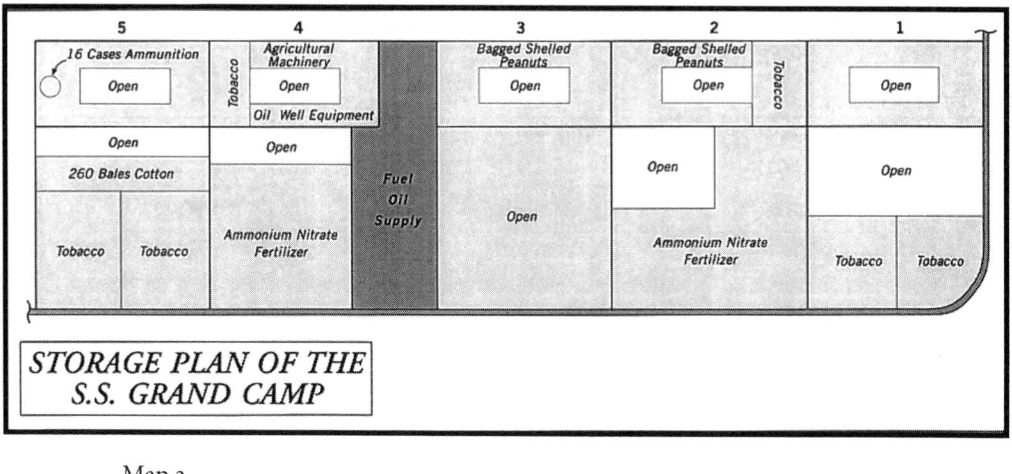

STORAGE PLAN OF THE S.S. GRAND CAMP

Map 3

Mrs. Willie-Dee Criss was looking out of her bedroom window when the ship exploded. She remembers seeing a wall of fire, "curtain-like streaks of reds, yellow, and white heat with black lacy edges on top."[30] John Hill, whose home was about two miles from the docks, jumped out of bed in time to see debris and bodies flying through the air as well as the dark ridge of a shock wave which soon hit his house. At that same moment, sixty miles away in Port Arthur, Ivy Steward Deckard's house vibrated, the windows rattled, and pictures danced on the wall. Frank Woodyard, a mate on the *Wilson B. Keene*, moored next to the *High Flyer* in the adjoining slip, remembered something resembling a heavy bombardment, "a dull rumble and rolling and cracking and crashing" which lasted about ten seconds.[31] Others recalled another loud explosion immediately after the initial blast, possibly a secondary blast at Monsanto or the results of a red-hot piece of shrapnel penetrating an empty 25,000-barrel spheroid tank at the Sid Richardson Refining Company, igniting a residual hydrocarbon-air mixture. A giant column of black smoke billowed 2,000 feet into the air, accompanied by flaming balls of sisal twine and cotton in the cargo. Soon thereafter, thousands of pieces of shrapnel of various sizes which had been the ship showered down upon port and town; according to J. W. Bradford, a guard at the Carbide and Carbon Chemical Company, this went on for several minutes. The first thought that flashed across the minds of some was that the Russians had dropped an atomic bomb.

The crowd of people and the explosion's fury formed a lethal combination of awesome proportions. Blast overpressure and shrapnel wreaked sudden death and terrible injury upon everyone around the North Slip. "In the twinkling of an eye," to use Ivy Steward Deckard's analogy, at least a thousand persons were transformed from healthy, active people into limp, dismembered dead or agonized injured. About a hundred known to be at the docks simply disappeared, later to be listed as missing and presumed dead, among them the twenty-seven volunteer firemen and members of Republic Oil's fire-fighting unit. Valuable, efficient commercial buildings and technology suddenly became ruins of twisted metal and burning rubble. Secondary effects of the blast were scarcely less lethal. A huge wall of water was forced up and out of the slip, creating a tidal wave 15 feet high which swept over the immediate area. It tossed a 150-foot steel

Blast victim at the docks. National Archives.

hydrochloric acid barge weighing 30 tons approximately 100 feet inland, where it came to rest on the stern frame of the *Grandcamp*. Many of the onlookers actually died from drowning when this rush of water swept over them as they lay stunned on the ground or caught them as it receded, carrying them into the turning basin.

Most of the deaths and serious injuries occurred within 1,000 feet of Dock O. There were fewer casualties 1,000 to 2,000 feet from the epicenter because this area was relatively open. From 2,000 to 4,000 feet were about twenty-five blocks of residences; many persons here were lacerated by flying glass or otherwise injured by collapsing walls and roofs. Miraculously, 112 persons standing within 500 feet of the *Grandcamp* survived.[32] One was officer W. A. Reeves, who had set up a roadblock at the end of the slip. Reeves was knocked approximately 600 feet into a ditch filled with oil and water and almost drowned before he was pulled out. Jim Newlin was about 100 feet from the stern of the *Grandcamp*; when he regained consciousness, he found himself about a mile away in a hole filled with four feet of oily water. All his clothes had been blown off except for his shirt collar, shoes, and belt lining. Another survivor was Willie Churchill, a longshoreman, who was walking toward one of the fire trucks when the blast occurred. He heard only a puff and was picked up and dropped about a block and a half away, bruised and cut. Like those of the other survivors close to the *Grandcamp*, both his eardrums were ruptured. H. O. Wray, who worked in the Terminal Railway office, likewise had his eardrums ruptured. Testifying about his experience immediately following the blast, he remembered: "Frankly, I thought it was the Resurrection Day. There was nothing I would recognize in the whole performance, and while I stood there I wondered why these and other corpses didn't get up. . . . It struck me as rather peculiar, knowing that I was a white man, that I would be amongst so many colored people. I did not realize until several hours afterward . . . that I was the same color myself."[33]

It is impossible to estimate how much damage this explosion inflicted upon surrounding buildings and other property because an assessment was not carried out before the *High Flyer* blew up. Subsequent investigations estimated that blast overpressure converted most buildings within 1,500 feet of Pier O to rubble, while all other structures within this radius were heavily damaged by heavy shrapnel and fire. The outer end of Warehouse O was leveled, and even the foundations of the last third of this pier were blown away. At the land end of the slip, an office of a molasses company and three of its tanks collapsed, releasing thousands of barrels of this thick, heavy substance onto the surrounding area. The large Seatrain crane used to transfer railway cars on and off ships, located on the opposite side of the North Slip in front of the Monsanto plant, was heavily damaged but remained standing. The remainder of the immediate area, including the Terminal Railway switching yard, was a complete shambles, strewn with debris from wrecked automobiles, boxcars, and building rubble. Pieces of the ship, weighing from a few ounces to several tons, were thickly strewn upon the ground, about three to four feet apart. Heavy railway tank cars located two blocks from the explosion were picked up and thrown at least fifty feet. Warehouse A, which was made of reinforced concrete and shielded by Warehouse O, was only partially destroyed, and Pier B, where the *Wilson B. Keene* was moored, was still standing, as were the grain elevators located at the head of this slip. Months later, while clearing the turning basin, the Army Corps of Engineers dredged up pieces of bent and twisted steel plate from the *Grandcamp* located as far as 600 feet from the end of Pier O and weighing as much as 13 tons.

Located on the opposite side of the North Slip from the *Grandcamp*, the Monsanto plant experienced the full force of the blast, which destroyed or damaged beyond repair approximately three-quarters of this $20 million facility. The dock warehouse and a part of the polystyrene building, 350 feet distant, were obliterated as the walls and roof were torn away. The steam plant and power house were flattened. Partitions were demolished and windows blown out in the office and service buildings. The heat of the explosion ignited benzol, propane, and ethyl benzene in pipes and storage tanks, starting fires that raged for two days. Of some 451 company personnel and 123 contract workers present, 154 of the former and almost 80 of the latter were killed outright or died of injuries, and another two hundred persons were injured.[34] H. K. Eckert, the plant manager, had glass shards blown into his head but survived. Dr. William Lane was in the laboratory at the time of the blast. The ceiling fell down on him, and his eyes, ears, nose, and throat were filled with asbestos. The plant fire team, who were laying hose lines on the Monsanto side of the North Slip, were not so lucky; all perished in the blast, and their 500-gallon pumper was picked up and thrown 200 feet.

Although relatively few persons were killed outright beyond the immediate blast area, falling shrapnel and residues of the cargo vastly extended the radius of harm. Pieces of the ship, some weighing as much as 2 tons but most around 20 pounds, showered down upon the town and neighboring refineries and tank farms. A 20-ton section of the *Grandcamp* was thrown 2,500 feet from the blast, shearing off several railroad tracks as it landed. Elements of the keel, measuring nearly 40 feet long, were thrown almost a mile. Storage tanks and processing facilities at Atlantic Pipeline, Stone Oil, and Republic Oil were punctured or damaged. Only seven oil tanks caught fire, but these produced a large cloud of heavy black smoke which began drifting across the waterfront. Fires broke out at the south end of the turning basin, where exposed pipelines leading from storage tanks to the oil-loading docks were ruptured by concussion or missiles, but there appear to have been few synergistic effects at the time. The fact that a 30-foot drill stem weighing 2,700 pounds was hurled 13,000 feet and that the *Grandcamp*'s 1.5-ton anchor was found 2 miles away embedded 10 feet in the ground at the Pan American refinery graphically demonstrates the incredible power of the explosion.[35] Residences and stores in the business district were also heavily damaged. Nine of ten homes in Texas City suffered significant damage, some of which may been inflicted when the *High Flyer* blew up. Houses closest to the docks, roughly south of Texas Avenue and east of 6th Street, being of light frame construction, were thoroughly devastated. Blast overpressure ruptured water pipes within some homes. Buildings farther away in the downtown business area also collapsed as walls and roofs caved in. Afterward, an insurance agent could not recall seeing a single unbroken plate glass window.

The force of the ship explosion had another important effect. It tore the *High Flyer* from its moorings at Pier A, allowing it to drift across the Main Slip and lodge against the *Keene*. Hatch covers were blown off, and superstructure on the port side and booms and working gear were completely wrecked. Even the main deck plates were warped and buckled. Although only one of the crew was killed, many were injured, and almost everyone was in a state of shock. According to the *Keene*'s chief mate, crew members were so thoroughly covered with mud and slime that they were difficult to recognize. Even so, some of the *High Flyer*'s crew ran a line from the stern to the *Keene*'s bow in order to secure their ship. Very quickly, visibility was reduced by thick, acrid black smoke coming from Pier O, while fumes from burning sulfur in Warehouse A made breathing difficult.

About thirty minutes after the *Grandcamp* blew up, a series of explosions along the waterfront occurred at short intervals, causing the *High Flyer* to surge. Lacking instructions from Lykes Brothers and believing tugs would not approach the ship and undertake the dangerous and difficult task of towing it away from the dock, shortly after 10:00 A.M., Captain Petermann ordered off the crew. They left the ship by going over to the *Keene* and climbing down a rope ladder onto the dock. There were no fires anywhere on the ship at the time, but Captain Petermann remembered being concerned that flying sparks might start one.[36] He and the crew then waded through a foot of oily water for a considerable distance before encountering a rescue party. They were taken by ambulance to a hospital in Galveston. Captain Petermann later testified that he asked his chief steward to telephone the Lykes Brothers office in Galveston with information as to his whereabouts.

As extensive as death and destruction were, the results could easily have been worse. The *Henry M. Dawes* was scheduled to take on naphtha at the Republic docks that day, but the master had made a last-minute decision to go back to Port Arthur for minor repairs.[37]

In the context of general complacency about an industrial disaster, pervasive indifference among officials about the explosive potential of the fertilizer, and lax safety standards, a fire which was probably started by a carelessly discarded cigarette inflicted a veritable holocaust on Texas City. A sudden explosion of unprecedented power from the marine environment visited its worst effects on the adjacent land area. It ripped through the heart of the industrial area and extended into the business district, killing hundreds, wounding several thousand, and bringing ruin to homes, stores, industries, and vital services. The sheer extent of the devastation was bad enough, but no preparations were in place to respond to its awesome destructiveness, thereby exacerbating human misery and property destruction as survivors began an extended struggle to sort out the confusion and havoc.

3

Chaos and Courage

OVERVIEW

Seldom is a disaster a total surprise. Usually, those who are
vulnerable anticipate the possibility of disaster. In Texas City,
vulnerability was all the greater because no preparations had
been made to cope with a major industrial emergency. Surprised and confounded by the horror of the moment, some
citizens fled in panic. But others faced danger resolutely,
extracting the wounded from the docks and providing first
aid. Law enforcement and medical and military personnel from
nearby towns converged on the scene quickly and provided
critical help in the early stages of the disaster. Even so, the initial response effort was much too disorganized and limited to
contend effectively with the intensive devastation caused by
the explosion of the *Grandcamp*.

At 9:12 a.m., police chief William Ladish was standing at the window of
his office at city hall, located at 6th Avenue and 6th Street, watching smoke
rise from the docks. His reaction to the explosion of the *Grandcamp* epit-
omizes the resolve which quickly replaced the initial fear experienced by
those in responsible positions. Although the building was at least 6,000
feet away from the North Slip, blast overpressure slammed Chief Ladish to
the floor, and glass shards from the shattered window inflicted minor cuts.
For some reason, he and the dispatcher scurried into the toilet, where they

remained for about a minute. Discovering that the police radio was knocked out, Ladish ran to the telephone exchange through glass-littered streets filling with dazed and bleeding people. There, he placed a call to the Houston Police Department, where he talked to Captain W. M. Simpson. Informing him that a ship had blown up with serious loss of life, he requested fifty officers and all available ambulances.

With no more detailed information than this, Captain Simpson began recruiting law enforcement and medical assistance throughout southeast Texas. Actually, this was not the first plea for help from Texas City; telephone company supervisors had already placed distress calls to the National Guard in Houston, the John Sealy Hospital in Galveston, and the city hospital in neighboring Goose Creek. Returning to city hall, Chief Ladish ordered some of his officers to establish roadblocks and seal off the dock area. He sent the few remaining men of his seventeen-man department, soon supplemented by former auxiliaries of the wartime emergency force, to the docks with no more direction than advice to assist in any way possible and use anybody who would help.

Mayor Curtis Trahan had also been aware that a ship was on fire. After leaving one of his children at school, he drove toward the waterfront, where he encountered a policeman at a roadblock who informed him that the situation was under control. Although he cannot remember why, he then went to the municipal utility barns near the Republic Oil refinery. He was in his car talking to one of the employees when the *Grandcamp* exploded. "It seemed as though the earth was trembling and there was an optical illusion of some sort that looked almost like these barns were spreading out and would collapse," he later related.[1] He immediately started toward city hall, stopping to calm a group of frightened women, taking some children to their homes, and ascertaining that his wife was safe. Throughout the remainder of the 16th, he and Chief Ladish struggled against heavy odds to make headway against the chaos suddenly thrust upon the town. Among his first actions was to issue a disaster declaration and order Dr. Clarence Quinn, the city health officer, to set up first-aid stations. Assistant fire chief Fred Dowdy, who had been out of town that morning, quickly returned and assumed direction of a Fire Department which had just lost half its men and all its equipment.

The local Red Cross chapter chairman confronted much the same situation as municipal officials. Resisting the impulse to see about his family, he made his way to the telephone exchange to request help from the Galveston County chapter. Before very long he realized that his organization could not provide meaningful relief. In fact, hours were to pass before he or municipal officials obtained even a general idea of the challenge they confronted. There was little they could do in the meantime but improvise with whatever was at hand and wait for outside assistance.

Much of the very limited facilities useful for emergency operations was destroyed by the blast. The Fire Department had no equipment, and even the municipal building, the focal point of direction for relief efforts, was in a shambles. This building housed the mayor, the chief of police, a courtroom, a jail, and an adjoining auditorium. Before 9:12, it could have been described as a new, modern structure, but not afterward. Although just over a mile from the North Slip, it was badly damaged by the explosion. Offices faced the harbor and received the brunt of the shock waves: casement windows were blown in with such force that their steel frames were torn from the walls, furniture was smashed, files were strewn about, and telephones were put out of order. Normal communications were disrupted, and electric power and water systems were severely damaged. Unable to obtain an accurate idea of the devastation and secondary hazards created by the ship explosion, municipal officials could not control recurring rumors of imminent explosions or toxic gas releases which continually fed the anxiety of already-distraught citizens. As an Army report bluntly stated:

> The Mayor of Texas City and his assistants were endeavoring valiantly to bring order out of this confusion, but were handicapped by the hysterical attitude of the citizens, the rumors and reports which spread rapidly among the crowds, the lack of communications within the city, the lack of any wherewithal with which to work, and the ruined or semi-ruined condition of public buildings which prevented the various officials from carrying out their functions.[2]

It was almost impossible to obtain an accurate idea of the situation at the docks; fires burned fiercely among buildings, and heavy black smoke from burning tanks of crude oil shrouded the area from view. From time to time, explosions erupted at the Monsanto plant and at tank farms and refineries. Throughout the day, rescue and relief efforts were never organized well enough to assess damage accurately and map a plan of action for effective use of the outside help that was pouring into town. A Bureau of Mines official who arrived about 9:30 P.M., some twelve hours after the blast, found no semblance of coordination in response activities.[3] John Hill, who labored throughout the day and all night organizing committees to carry on vital activities, later remembered: "Had such an organization existed at the time of the disaster in Texas City, the control of the community in the first terrifying hours succeeding the disaster would have been greatly facilitated."[4]

Perhaps it was remarkable that anything meaningful was accomplished at all. Even without direction, immediate and obvious requirements, such as search and rescue, medical treatment, and traffic control, were spontaneously initiated. Units of the armed forces from nearby bases soon arrived and rendered useful service, as did veterans, who drew upon training and experience obtained in the recent war. Nevertheless, other critical activities were carried out poorly or not at all, particularly those of a less obvious nature and those that required coordinated action among several organizations.

Groping for Order

No sooner had shock waves dissipated and shrapnel ceased falling from the sky than a curious pattern of activity developed amid the ensuing chaos. Some persons immediately converged on the waterfront to search for relatives and friends. George Gill cobbled together about a dozen volunteers from the Carbide and Carbon Chemical Company plant at the edge of town, stripped the plant fire truck of all available extinguishing equipment and gas masks, and rushed to the scene. Stopping close behind the Monsanto property, they began assisting wounded emerging from the maelstrom. Frances Alexander, who worked at a laundry, quickly set out toward the inferno in search of her twelve-year-old brother Gilbert, who had gone to the docks to watch the fire. "I ran toward the explosion, looking for my little brother," she later related. "I saw Gilbert and I could hardly recognize him. One eye was hanging out, and

Damage to houses near the docks. Moore Memorial Public Library, Texas City.

one side of his face was out of shape. One arm was broken. I started crying. Gilbert said, 'Sister, don't cry; I'll be alright. Go see about mother and the others.'"[5]

Many more, particularly those who were between 1,000 and 2,000 feet from the North Slip at the time of the blast, took the opposite course of action. A man who had driven behind the Monsanto plant shortly before the explosion remembered seeing a substantial number of sightseers on the road; when he walked back down the same road a few minutes later, it was entirely deserted. Mayor Trahan recalled a "sea of people" rushing away from the residences south of Texas Avenue. After extricating themselves from their collapsed dwellings, those living in the flimsy houses closest to the waterfront and refineries fled precipitously on foot or in automobiles until they were well out of town.

Made of materials used in army barracks, Booker T. Washington Elementary was the nearest school to the docks. Lynn Ellison, now a member of the city council and then a first-grader, remembers that his teacher pushed children out the window because the walls separating the classroom and the hall had collapsed. Rosa Lee Curry had just seated her fifth-grade class when the building shook and something crashed through the roof. Quickly she herded her children out of the building and headed northwest, away from danger. Knocked down in a ditch, she was pulled out by one of her pupils. After being picked up by the driver of a flatbed truck, teacher and children were taken to Barbour's Chapel and eventually to nearby Dickinson.

The Reverend F. M. Johnson, pastor of the First Baptist Church and one of the few black persons with an automobile, spent much of the day transporting refugees out of harm's way. Among those who fled in their automobiles was Henry Criss, a plumber who had worked at the polystyrene plant at Monsanto. Although slightly injured by flying glass from the bedroom window and under the impression that the Monsanto plant had exploded, he quickly piled his family into the car and went out "a back way" to Houston. "We never looked back. Just came on to Houston as fast as we could."[6] Others lacked even this sense of purpose; a mortuary student who arrived in Texas City during the afternoon noticed that "men, women, and children were dazed, walking away from town on the highway, not knowing where they were going and seemingly not caring."[7] The headlong flight of hundreds of persons completely out of town meant that several days passed before some families were reunited; in the meantime, many assumed that their loved ones had been killed. All too often, that was true.

Those who happened to be in the business district and outlying residential areas were just as surprised as anyone. Most had not even realized that a ship was on fire. A few were killed or seriously injured when they were struck by falling shrapnel, but scores suffered minor injuries as roofs, ceilings, and walls collapsed, windowpanes were transformed into glass shards, and furniture and decorations were flung about rooms. Some merchants who had just started the business day rushed home to their families, leaving cash registers open and goods unguarded. Plate glass windows were shattered, and shards littered sidewalks and streets. "I can't remember anything except walking on glass," stated Mary Garcia Parker as she and her mother searched for their father and husband, Carlos.[8] Coy Poe thought that the town looked as though it had been hit by an earthquake, except that huge pieces of steel were strewn all over the place, and cotton from the *Grandcamp*'s cargo was so thick upon the ground in some areas that it resembled snow.

Schoolchildren quickly added to the confusion. Although all schools except Booker T. Washington Elementary were at least 4,000 feet from the North Slip, blast overpressure shattered windows, and ceiling partitions and beams collapsed in buildings. No one received serious injuries, but a considerable number of children were cut or otherwise hurt. As pupils were evacuated from damaged buildings, a dilemma quickly emerged which could easily have been anticipated if any forethought had been given to a disaster of this kind: wanting to find their parents or needing first aid, some children broke away or were released by teachers and ran toward home or to one of the three clinics in town. Confusion increased as parents converged on the schools, frantically seeking their children. Being much closer, children and adults with minor injuries quickly overran the clinics before the more seriously hurt were brought in from the dock area. Missing children remained a problem throughout the day, complicated by well-meaning adults who sometimes took them to neighboring towns to care for them. An entry in the police radio log at 6:25 P.M. indicates that the mayor instructed that all lost children were to be taken to the Mainland building, located in the business district.

Survivors near the waterfront had no idea what had happened and only thought to help friends and flee the scene. Ruptured eardrums rendered almost everyone temporarily deaf and knocked many unconscious; others lay trapped under collapsed ceilings and equipment. Most of the injured were so uniformly covered with a heavy, tarry substance of large soot particles, droplets of oil, and heavy chemicals that their race could not immediately be determined when they were taken to hospitals. To Harvey Williams, the Pan American Refining Corporation safety engineer who had left the docks in search of gas masks, survivors filtering out of the dock area presented a ghastly sight: "Grotesque figures they were, black with fuel oil and smoke, and red with their own and comrades' blood, walking with arms broken and dangling, or crawling with mangled legs and feet."[9] "Everything was utter bedlam," recalled Sally Wehmeyer. "Cars were being driven on the wrong side of the street; men, women and children were running about in their night clothes, some screaming and crying and others just standing with blank, stunned looks on their faces."[10] At the devastated Monsanto plant, a young engineer named Fred Grissom was blinded by blood streaming from wounds on his head. As he stumbled through the wreckage, he encountered another man whose legs were broken. By virtue of the other man's sight and his mobility, both were able to escape from the building. Leola Howell, who had been blown out of her car into a pool of oil on the company's parking lot, struggled through smoke so thick that she could see no more than a foot ahead. Like many of the dead, some survivors had their clothes almost completely blown off their bodies. Cut and bleeding, Carl Baker was able to hobble out of the wreckage of the Terminal Railway office. Everyone seemed to get excited when he walked to a gate. It was only when he was taken to a clinic and a nurse remarked upon his "pretty shorts" that he understood why; he was naked except for his shorts and belt.

Even so, while some fled and others wandered about the streets injured or dazed, individuals and small groups began a purposeful search for survivors. Employees from Pan American, Carbide and Carbon Chemical Company, Sid Richardson Refinery, and other facilities spared serious damage formed rescue teams and sent their firefighters to the docks. These were among the first of perhaps as many as five hundred persons who converged upon the waterfront over the next several days. In about two hours' time, officers and men from the Galveston district office of the Army Corps of Engineers arrived on the scene with trucks and heavy moving equipment. Workers plunged into the still-burning wreckage, first seeking the injured and leaving the dead for the time being. Those able to leave with assistance or on their own were quickly gathered up, but fires and rubble made it difficult to locate and extract persons trapped in buildings and automobiles. Getting so many wounded to medical treatment was a challenge; flatbed trucks, automobiles, buses, and, as the mayor later said, "anything else that would roll on wheels" were used for this purpose.

Initially, the only sources of emergency medical treatment were the three clinics and ten doctors in Texas City. One of the latter, Dr. Gerhard Manske, had been slammed to the ground by the ship's blast. Picking himself up, he ran to his office at the Beeler-Manske Clinic and found his colleague dazedly picking up instruments. Within five minutes, this office and those of Danforth and Tidwell-Schmidt as well were inundated by a wave of injured, most of whom had cuts inflicted by flying glass. Those with serious wounds soon began arriving in such numbers that patients were placed on reception room floors, in corridors, and eventually outside on lawns and sidewalks. Because clinics were initially without electricity, gas, or water, little could be done for patients beyond treatment to prevent shock. "Our limited facilities for large numbers of patients made the task almost hopeless," Dr. Clarence Quinn later remembered.[11] Footing itself became difficult as more and more blood streaked the floor of the crowded rooms. Ordinary citizens also spontaneously provided first aid. Informed that children needed first aid, one merchant ran into a pharmacy seeking medical supplies; with the owner's permission, he and a companion scooped up items scattered on the floor and set up an aid station in a vacant lot. This sort of generosity was typical of merchants; others gave away stocks of blankets and bedding as well as medical supplies.

Systematic efforts to extract the wounded at the waterfront were slow to get under way. By noon, having heard of the tragedy in dramatic broadcasts on both local and national radio, individuals and groups of volunteers began filtering in from Houston and neighboring towns. Many gravitated to the docks to join the small groups that had undertaken the dangerous tasks of search and rescue. Those most familiar with a particular location or equipment useful in extricating the injured assumed leadership roles. "At first there were little groups doing separate jobs," stated John Hill, who became the mayor's chief assistant and oversaw efforts to coordinate municipal activities. "There were many standing around who seemed to want to help but did not know what to do. They were glad to follow any reasonable instructions. These separate crews started out working on their own, but they began to get together when they would come up against big jobs."[12] By all accounts, much of the cohesiveness of these

Volunteers removing victim from the dock area. Moore Memorial Public Library, Texas City.

groups under the stress of the moment derived from the discipline and experience of those who had served in the armed forces during World War II. Some who sought to help could contribute very little because they lacked equipment, including simple items like work gloves.

Work at the docks was dangerous and difficult. Tanks and pipelines containing natural gas, butane, and other flammable and highly explosive products posed a constant threat. For the first couple of hours, those searching for the injured around the Monsanto plant had to contend with oily water reaching up to their knees. From time to time, rescue workers were physically driven back by noxious fumes or fires or pulled back from the waterfront because of rumors of impending explosions. Visibility was seriously impaired by smoke from fires along the waterfront and from burning crude oil tanks at the Stone Oil Company; the commander of the Coast Guard buoy tender *Iris*, who arrived at the turning basin shortly before 10:00 A.M., could not fix the precise location of fires because thick smoke clouds continually rolled across the waterfront. Since initial rescue efforts focused upon Monsanto and the North Slip, this meant no one had a clear idea of the situation farther down at the Main Slip, where the abandoned *High Flyer* and *Keene* lay tangled together.[13] Leaving the *High Flyer* about 10:00, Captain Petermann found that familiar landmarks had been completely obliterated: "All we had to guide ourselves by was the Grain Elevator. . . . There was anywhere from six to eighteen inches of water all over the dock area. The docks themselves were almost completely demolished and we had to pick our way through."[14] Visibility in the area was so poor that Swede Sandberg, who knew the waterfront intimately and gained widespread respect for his tireless efforts on the 16th, was under the impression for most of the day that the *High Flyer* was no longer in port.

Fires, smoke, and the threat of explosions were only part of the trouble. Before the *Grandcamp* blew up, vehicular access to the docks was by means of a single shell-topped road and a side road leading through a tank farm. The former was now blocked by the *Longhorn*, a 150-foot-long hydrochloric acid barge which had been thrust up on land, and by the twisted frame of a two-story steel conveyor thrust between Warehouse B and the grain elevator. As a result, rescue efforts were funneled into an area heavily cluttered with wreckage of employee automobiles, building rubble, and large pieces of the *Grandcamp*. At another point, workers from

Carbide and Carbon Chemical Company had to round up dump trucks and bulldozers and construct a small causeway across a large drainage ditch in order to reach some wounded persons.

Throughout 16 April blazing fires seriously impeded search and rescue efforts at the docks. Many were fed by oil spilled from broken pipes leading to tank farms and others by the scattered remains of jute bagging and cotton scattered around Warehouses B and C. Burning sulfur in Warehouse A compelled rescuers groping through the black smoke to wear gas masks. The heat from flaming tanks of benzol, propane, and ethyl benzol, sufficiently hot to melt steel supports, prevented anyone from approaching most of the Monsanto complex. Early on, Captain Glen Rose of the Texas Highway Patrol tried to reach employees in the area but could only get close enough to hear their screams as they burned to death. The loss of Texas City's fire trucks made little practical difference to fire fighting at the time; although elevated water tanks at the docks were intact, hydrants and pipes had been split open by the blast, and pump houses at the Terminal Railway and at Monsanto had been leveled. Having experienced approximately two hundred breaks, municipal water lines were closed to conserve the existing supply. At any rate, the municipal tank could not operate again until emergency power was provided. Even though foamite for fighting oil fires was available, it was useless without water in which to dissolve it. Units from outlying refineries, other towns, and military bases made some progress putting out fires during the afternoon and early evening, but for the most part these fires were behind the immediate waterfront area. Effective suppression was not to begin until the 18th, when the single shell-topped road was sufficiently cleared of rubble to allow heavy fire-fighting equipment supplied by the Houston Fire Department to reach the docks and draw water from the turning basin.

Help from Outside

Shock waves and billowing black smoke affected other Texas cities in the area. In Galveston, large buildings swayed precariously, causing several thousand persons to run into the streets to escape the effects of what they thought was an earthquake, only to be driven back inside by falling debris. Buildings also vibrated in Baytown, fifteen miles to the north.

Automobile drivers in the area who immediately diverted from their destinations to head for Texas City were guided by a smoke column which quickly climbed almost 4,000 feet into the sky. Cliff Wattam, a crew member on the Coast Guard buoy tender *Iris*, remembered "the damnedest explosion I ever heard" just as his boat was casting off from the dock at Pelican Island in Galveston.[15] Soon, heavy black smoke and murk arrived, carried by the brisk north wind. Patty Parish Hurt, a student at Galveston's Ball High School, later recalled that after she and her fellow students were evacuated to a park across the street, everyone's clothing became spattered with minute black dots.[16] For all his concern about what had happened at Texas City, Wattam was miffed because this substance quickly ruined his morning's effort scrubbing the white paintwork on the bridge of the *Iris*.

Spurred by lurid radio broadcasts of the disaster, residents of neighboring towns put aside their ordinary activities and threw themselves into a concerted, extensive outpouring of assistance. The circle of help quickly expanded to include Houston and all of southeast Texas, then Dallas, San Antonio, and much of the rest of the nation. Because information trickled in slowly, initial reports were of limited use to those who were available to help. Chief Ladish's call to the Houston Police Department stated simply that a ship had blown up and requested all possible assistance. An oil company worker telephoned the editor of the *Houston Press* and informed him that the Monsanto plant had blown up and a thousand people were dead. Headquartered at San Antonio, the Fourth Army, which assumed a key role in managing initial response activities, learned of the event when notified by an FBI agent in Houston who had received a report from a newspaper that a ship had blown up with tremendous force at either Texas City or Galveston.[17] Although state highway patrolmen converged on the area as soon as they heard the news on their car radios, more than thirty minutes passed before the Department of Public Safety understood that not only had a ship exploded but the industrial area of Texas City had been devastated as well.

Galveston provided Texas City's most pressing need—medical assistance. Using elements of its hurricane relief plan, the entire medical community quickly mobilized for this task. It so happened that the chairman of the county's Red Cross disaster medical aid committee had witnessed the explosion from his office window in Galveston. He immediately alerted the city's three hospitals and the Red Cross chapter to prepare for a large number of casualties. At John Sealy, a teaching hospital, classes were dismissed, and faculty, many of whom had served as military surgeons during World War II and were familiar with mass casualty situations, organized themselves and their students into teams to handle the expected influx. Within thirty minutes, Galveston ambulances and buses —some taken off their regular routes—congregated at the hospitals to pick up doctors and nurses. Among the first to depart was a group of high school students, because the principal of Ball High had been asked to provide stretcher bearers: "In about ten minutes I had pulled out 50 husky boys and started them towards the hospital to get on the ambulances."[18] The convoy's movement was facilitated by self-appointed traffic wardens who took up positions and stopped vehicles from crossing the four-lane highway leading to the causeway. This convoy and a contingent of almost fifty vehicles from Fort Crockett, the local army base, formed part of a stream of ambulances, fire trucks, taxis, heavy construction equipment, buses, military vehicles, and private cars which spontaneously converged upon Texas City. Two Army first-aid and clearing station teams which arrived at Texas City within an hour of the *Grandcamp* explosion were in the vanguard of an estimated 1,250 doctors, nurses, and first-aid workers from military services, the Red Cross, or private practice who either worked at the scene of the disaster or treated the wounded at hospitals in Galveston and other nearby communities.

As the Galveston County chapter of the Red Cross mobilized to meet the demands of a catastrophe quite different from a hurricane, it appealed to other chapters for penicillin, tetanus serum, blood, and surgical supplies. The city's three major hospitals—John Sealy, St. Mary's Infirmary, and the Marine Hospital—immediately discharged as many people as possible to make room for expected casualties. Off-duty nurses and Grey Ladies reported without being summoned, and merchants sent additional beds, linens, surgical supplies, drugs, and clothing, often without even being asked. Additional hospital space was created when Mrs. James Tompkins, wife of the senior Lykes Brothers official in Galveston, broke into shuttered Fort Crockett Army Hospital with the assistance of a neighbor and began making preparations to receive casualties. These ladies were soon joined by Colonel E. M. Mitchell, the Army ROTC representative at Ball High School, who took charge of this effort on behalf of the commanding officer at Fort Crockett. Within a few hours, elements of the Thirty-second Medical Battalion dispatched from Fourth Army headquarters arrived to staff this facility.

The first wave of doctors and nurses arrived in Texas City about the time the town's three clinics began filling up with the seriously injured extracted from the waterfront area. Some medical personnel, particularly those from nearby military bases, went directly to the waterfront while others gravitated to one of the clinics or a first-aid station. The situation was a grim one; clinics were hampered by an absence of water, electricity, and natural gas as well as by blast damage. Even as more doctors and nurses arrived and plasma, whole blood, and other items were delivered, a steady influx of the severely wounded over the next several hours quickly overtaxed capacity. So many injured collected outside the clinics that the city auditorium and high school gym were pressed into service as first-aid stations. Definitive care could not be administered to the seriously injured; they were treated for shock with morphine and blood plasma to keep them alive until they could be transferred to a hospital. "We were handicapped no end by not being able to hospitalize these cases in Texas City," wrote Dr. Clarence Quinn, the medical coordinator. "Many cases were marked for evacuation which could have been hospitalized here in Texas City had we the hospital space."[19] Even with stocks brought from Galveston, the supply of available blood plasma was exhausted by noon. Nonetheless, only about fifty serious cases sent to the hospital subsequently died. The larger number of less seriously injured received first aid and were told to return in two or three hours for final treatment.

Unprecedented casualties created other problems. One was transportation of seriously wounded persons to hospitals. Texas City had only two serviceable ambulances, since the others had been at the docks when the *Grandcamp* exploded. Initially, two buses were loaded with patients and sent to Galveston, but as more collected at clinics and aid stations, trucks, taxis, and private cars were pressed into service. Of 852 severely injured

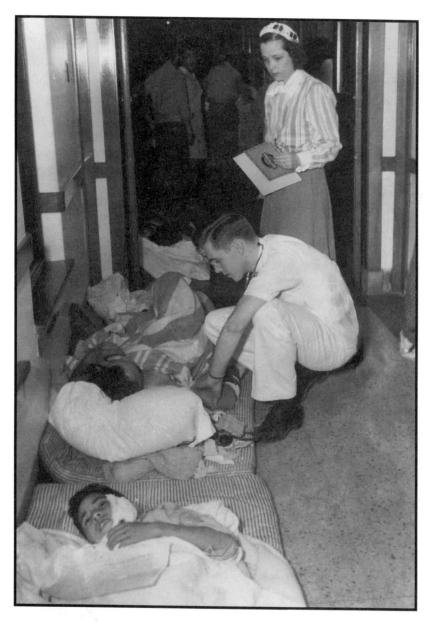

Casualties in a Galveston hospital, 16 April. Moore Memorial
Public Library, Texas City.

persons, about two-thirds were taken to Galveston and many of the rest to Houston, although twenty-one different hospitals received casualties.[20]

Galveston's three hospitals were in the midst of preparations when the first wounded arrived. Some came on stretchers conveyed by ambulances or buses, but others were carried in on planks and doors. The hospitals, particularly John Sealy, became beehives of activity. Casualties were brought to an outpatient clinic, where a determination was made as to whether they should go immediately to an operating room or a shock ward. Those not requiring surgery were sent to Fort Crockett for further observation. Ten operating teams of three or four doctors and nurses, later supplemented by military surgeons and nurses, worked in nonstop shifts of forty-eight hours. By 11:00, casualties overflowed into corridors and onto mattresses on the floor. Ken DeMaet, who brought his bloodied wife there, remembers waiting several hours at John Sealy Hospital before a doctor was able to treat her wounds. According to a nurse at St. Mary's Infirmary, the wounded were stacked "like floor tiles." After surgical gowns became soaked with blood, doctors continued to work bare-chested over the operating tables. All told, slightly more than two hundred army doctors, nurses, and orderlies served either at one of the three hospitals or at Fort Crockett. Victims who had been outside when the *Grandcamp* blew up suffered from blast effects and compound injuries from flying missiles. Those inside buildings tended to have multiple perforating wounds, especially glass splinters, and crushing injuries. Almost all casualties had perforated eardrums and fractures. Those who had been near the North Slip had a thick black coating of oil and soot and wounds "contaminated with every imaginable type of debris: dirt, sand, molasses, oyster shell, particles of wood, glass, grease, oil, etc."[21]

As the injured were evacuated and volunteers rushed in to help, Texas City came to resemble an anthill whose surface has just been scraped off—there was much scurrying back and forth. There were but three roads into town, only one of which was a major thoroughfare. At first, vehicles were able to move at a high rate of speed, but by 11:00 these vital arteries had begun to fill with ambulances, fire trucks, heavy equipment, and private vehicles. When Fred Aves, a physician employed by Lykes Brothers, went to Texas City on company business in midmorning, he found traffic and bedlam along the roads "terrific." By early afternoon, gridlock threatened as curiosity seekers and relatives of citizens descended upon the town and substantial numbers of residents began leaving, frightened by recurring rumors of impending explosions and toxic gas releases. Conditions were little better within Texas City itself as vehicles prowled outside the dock area looking for the wounded or sped in all directions as their drivers tried to make themselves useful.

Fortunately, state and local law enforcement officers also converged upon Texas City and helped establish access control to the town and patrols to preserve order and prevent looting. Early on, police auxiliaries and plant guards set up roadblocks on their own initiative but with only par-

Armored car roadblock, 17 April. AP Wide World Photos.

tial success, since virtually anyone who claimed to have official business was allowed to proceed into town. At the mayor's request, in little more than an hour after the *Grandcamp* exploded, Lieutenant Colonel George McLean, Jr., of the Texas State Guard established headquarters at Kohfeldt High School and mustered a twenty-man detachment of the Forty-ninth Battalion to assist in maintaining order within the municipality.[22] Although assignments were made haphazardly at first, a measure of coordination evolved because outside law enforcement officers reported to Chief Ladish and organized themselves at his direction. A rough division of responsibility was worked out, allocating control of roadblocks to the ranking Texas Highway Patrol officer, maintenance of internal order to the Texas Rangers, and control of an area extending half a mile from the docks to a Houston police captain. This arrangement, confirmed at a meeting called by the mayor at 1:00 P.M., facilitated integration of sizable contingents of local police from Houston, Galveston, and San Antonio, as well as air force personnel from Ellington Field, southeast of Houston. Given this circumstance, Chief Ladish was able to dissuade Mayor Trahan from requesting a declaration of martial law. When Governor Buford Jester arrived in the late afternoon, he declared only a state of emergency, allowing the municipality to retain control of law and order.

Local police and sheriff's departments, state police, and state guard and regular military units eventually contributed about fifteen hundred personnel to assist in access control and to deter looting.[23] Volunteers from other towns who arrived in the early afternoon do not recall having any difficulty entering the town, but gradually roadblocks began to take effect. Although residents living near the docks reported some losses, and valuables appear to have been removed from bodies, looting was minimal. State troopers screened out the curious from the helpful and inspected departing cars for stolen goods. According to Colonel Homer Garrison, director of public safety for the state of Texas, by nightfall real progress had been made in removing nonessential persons who had arrived earlier in the day.[24] Roadblocks did not clear up traffic congestion, however. Possibly as a result of the explosion of the *High Flyer,* when the assistant director of disaster relief for the Red Cross started out from Galveston for Texas City early on the morning of the 17th, the main road was so jammed with vehicles that he was obliged to get into town over back roads, past blazing oil tanks and burning buildings.

Another kind of traffic congestion arose in the afternoon as a steady stream of planes began arriving at the town's small, privately owned airport with medical personnel and emergency supplies. Men stood in the middle of the runway directing traffic and then used a signal lamp as night fell until an emergency radio truck supplied by the Civil Aeronautics Agency was placed in service that evening. Larger military transport planes from various points in the United States carrying medical personnel and supplies were obliged to land at Galveston.

Despite the state of shock among the citizens and the confused circumstances prevailing during the first eight hours, several essential response activities were carried out quite well. This was especially true of medical assistance, access control, and maintenance of civil peace, because the groups involved possessed the skills and discipline appropriate for emergencies. Other essential functions, particularly communications, lacked these advantages and required much more time to put into operation. These still remained sporadic and uncertain both within Texas City and with the outside world until the next morning, when the night-long efforts of the mayor and John Hill to bring a measure of overall organization to response began to bear fruit. Until 10:00 A.M. on the 17th, no one had an accurate idea of the overall situation, so officials were hard pressed to give form to the overall response, particularly since rumors of impending explosions or toxic gas releases continued to circulate through the town.

4

Struggling for Order

OVERVIEW

As the injured were collected and treated, additional volunteers and equipment arrived at Texas City, giving form to medical assistance and search and rescue efforts. But shortages of
equipment, supplies, and skilled personnel soon developed, and
response remained piecemeal and uncoordinated. Fragmented,
sporadic communications were the root cause of all sorts of
problems. Rumors erupted periodically, disrupting search and
rescue at the docks and inspiring periodic flights of fearful citizens from town. Because no one really understood the overall
situation, the outpouring of spontaneous assistance from neighboring towns, the state, and the nation was not used well during the immediate aftermath of the *Grandcamp*'s explosion.

As the morning wore on, a flow of doctors, nurses, firefighters, bulldozers, field cranes, gas masks, and urgently needed medical supplies
began arriving. Among the first groups to arrive was an Air Force unit of
fourteen men with a fire truck and an ambulance from Ellington Field,
southeast of Houston. This team was instrumental in bringing a fire at
Republic Oil under control in the early afternoon. The naval air station at
nearby Hitchcock sent ambulances and medical personnel, and the naval
hospital at Houston provided seventy doctors and nurses, setting up an
emergency field hospital at the municipal auditorium in Houston. Spot
appeals broadcast on Houston radio stations started another fifty military

reserve doctors, pharmacist's mates, and corpsmen toward Texas City in midafternoon. By then, army and navy transport planes were winging toward Galveston from points around the United States carrying critical quantities of whole blood, plasma, penicillin, and tetanus serum. Already prepared for emergencies, utility companies responded efficiently. Except for the waterfront, electricity was restored by late afternoon, and trunk telephone lines were reconnected within a few hours. A meterman for the Houston Pipe Line Company alertly closed off the main block natural gas valve lines to the docks and then directed firefighters to fire plugs. More comprehensive measures followed when Division Superintendent J. M. Young of Houston Natural Gas arrived with a crew of about sixty. First, the company regulator station was cut, then the town was mapped off by sections, and crews were sent around to cut off meters at over three thousand residences. This precautionary measure had to be partially reversed about noon in order to restore the energy supply to clinics, cafés, and filling stations. The possibility that the municipality's water was contaminated because pressure had been lowered by ruptures in the lines prompted the authorities to advise citizens to boil their water. Tests soon revealed that contamination had not occurred.

Those who gravitated toward the docks and joined the search for the injured were mostly individual volunteers. As the afternoon wore on, rescuers gradually organized themselves into teams and, using heavy equipment donated by private contractors and the Army Corps of Engineers, slowly groped through heavy black smoke and chemical fumes, searching for more wounded. Salvation Army and Red Cross canteen trucks commenced operations, providing coffee and sandwiches for workers, and an army field kitchen set up in the business district began feeding both workers and citizens unable to provide meals for themselves.

Seeking Order

Spontaneous help was welcome, and heroic rescue did save lives, but coordinated action on a scale which might have measurably diminished the chaos failed to materialize throughout 16 April. Locating, treating, and transporting the large numbers of wounded was only part of the problem; numerous fires burned fiercely throughout the day, creating a continuous pall of smoke thick enough to obscure the sun. Fortunately for everyone except those at the docks, the northeast wind held, carrying flames and sparks away from town and out over Galveston Bay. Water for fighting fires along the docks and at petrochemical storage tanks was not available. Despite this, apparently no one thought to request the assistance of the Houston Fire Department, the largest in the area. Periodic rumors of impending explosions or poison gas releases obliged those at the waterfront to cease work and pull back, slowing the search for victims; alerts occurred about every two hours during the afternoon. The first, at 1:00 P.M., was of an impending explosion. The second at about 3:00 P.M. of a chlorine gas release was investigated and personally squelched by the mayor. When oil storage tanks at the Humble tank farm just south of the dock area caught fire from adjacent buildings and exploded about 5:00 P.M., another wave of panic swept through the town, spawning a rumor that the mayor had ordered a total evacuation. This rumor prompted the Army to prepare barracks for refugees at Fort Crockett in Galveston and at Ellington Field. Almost four hours elapsed before this report was discounted.[1]

Continuing fires and periodic explosions at petrochemical facilities near the docks thoroughly traumatized many citizens, leaving them vulnerable to lurid reports of impending dangers. Items were reported on commercial radio broadcasts or by sound trucks circulating throughout town. The cumulative impact of these reports prompted an even larger exodus of citizens during the afternoon than had occurred immediately after the *Grandcamp* blew up. Ultimately, about 40 percent of Texas City's residents left for a period of time, most to stay with relatives and friends, but some to refugee centers which were being set up at Ellington Field, nearby Camp Wallace, and the naval air station at Hitchcock. Others merely went to the north edge of town and waited until an "all clear" was given. But the net effect of such reports was to disturb townspeople throughout the day.

Available documents and recollections of survivors are neither sufficiently comprehensive nor detailed enough to trace precisely the nature and sequence of problems which confronted responders during 16 April. It is apparent that no one, not even municipal authorities, had more than the vaguest idea of the character of the damage; therefore, they were unable to use their limited resources to best effect. Even though the mayor hastily pulled together a small, informal staff of prominent citizens who had gravitated to city hall to help, without sufficient information to grasp the overall situation or a plan to guide their actions, to use his own words, he and those who worked with him "took on the job of more or less managing whatever situation happened to come up."[2] Arriving in the late afternoon, the advance party from Fourth Army headquarters found "confusion, disorder, and uncoordinated effort by the civilian population."[3] "There was no overall head during the entire operation at Texas City," reported a Coast Guard officer. "The mayor Mr. Trahan was supposed to have had this but from my observation we worked through the mayor [*sic*] office and Texas City police station and whoever appeared to be in authority at city hall."[4] "The lack of either a Disaster Plan or prearranged mutual aid was painfully evident," stated one postmortem. "No concerted effort toward fire suppression was achieved for at least 48 hours following the initial blast. . . . The wonderful examples of personal initiative and heroism which marked the early hours of rescue and salvage work lacked the necessary direction in many cases."[5] Given the severity and scale of harm and the complexities of coping with sudden devastation, valor and improvisation could not compensate for the fragmented character of response.

An absence of systematic record keeping added to the confusion of the day and seriously impaired the efforts of those struggling to gain some control of the situation, particularly so for tracing disposition of casualties. Nurses and volunteer first-aid workers did make notes about those they treated. But hurried, informal lists written down on scraps of paper at clinics and aid stations continually hampered efforts to track down individuals among several thousand killed and wounded, many of whom were taken to hospitals in other communities or on military bases. Distress was compounded because nothing was in place to track the whereabouts of over two thousand homeless persons and others who fled town. The local Red Cross chapter did manage to establish an inquiry office at the Chamber of Commerce around noon and communicated with the Galveston County chapter by means of teletype, but few knew of its presence, and it contributed little until the next day. KGBC, a Galveston commercial radio station, set up in the telephone company office and began broadcasting messages requesting information about missing persons, but it operated independently of the Red Cross. One important result was that as he valiantly struggled to organize a response, Mayor Trahan was constantly distracted by a stream of frantic persons who came to his office seeking information about relatives: "Always there were dozens of troubled people waiting to see him. He saw them all. He would talk with them, holding conferences while sitting on a cot in his office. No situation ever seemed to stump him. He was courteous but the moment a problem was solved he terminated the interview quickly."[6]

Much of the difficulty which responders encountered in sorting out the confusion on the 16th is attributable to difficulties in communications. This is often a problem in disaster response because disasters, by throwing large numbers of people into turmoil, generate a surge of urgent messages. Adequate communication amounts to more than having sufficient channels to handle the surge; operating under stress, responders must communicate accurately with others they often do not know in order to solve unfamiliar problems. Delays or counterproductive actions sometimes occur because decisions are made by persons who do not understand the larger context of the disaster. Evidence on this point is limited, but it is apparent that poor communications both within Texas City and with the outside seriously impaired efforts to utilize resources. As a result, a considerable period of time passed before even those working directly with the mayor obtained an accurate idea of damage and passed on requests for particular types of assistance. Because the police radio was knocked off the air for several hours, Chief Ladish was obliged to make his initial request for assistance to the Houston Police Department from the telephone office. Indeed, he was fortunate to be able to do so, for the main trunk line to LaMarque had been severed by the ship explosion and was not restored for a couple of hours. But follow-up calls were not placed to the Houston Police Department, and the captain who took Chief Ladish's initial call mobilized assistance throughout southeast Texas only with the knowledge that a ship had blown up. Only after a Coast Guard radio van set up operations next to city hall in midafternoon was something resembling a regular link available. Even with additional help from an amateur ham radio operator, neither the mayor nor Chief Ladish had an effective means of forwarding particular requests as needs developed.

Communications difficulties hampered delivery of assistance by the armed forces as well. Early on, an Army Corps of Engineers car with a two-way radio arrived at city hall, but its radio frequency and limited range meant that messages had to be directed to the engineer base at Fort Point, relayed by telephone to district headquarters, and then sent to their final destination. Army, Navy, and Coast Guard units were unable to communicate with each other because of different crystal settings on their radios. When the disaster team from the Fourth Army established headquarters at Fort Crockett on Galveston Island around 5:00 P.M., it could not establish contact with Texas City by telephone and was obliged to use a radio link consisting of a Coast Guard van and Corps of Engineers equipment for the remainder of the day. In order to augment what an army report characterized as "rather meager communication," high-powered mobile radio sets were obtained from Fourth Army headquarters. These went into operation about noon on the 17th but failed almost immediately. Not until the 18th were satisfactory communications established between Fort Crockett and Texas City.[7]

Major relief organizations fared no better. During the first two days of the disaster, the Fourth Army and the Red Cross operated largely in isolation, requesting help without reference to each other or to municipal authorities. This created gaps and overlaps in the delivery of assistance. For instance, when someone at the docks decided that a thousand gas masks were needed because chlorine leaks were possible, the request went to several different organizations: some ten to twelve thousand masks were sent from as far away as St. Louis. John Hill relied upon commercial radio to relay some appeals to citizens and police instructions, but many of those elsewhere who wanted to help could do little more than sympathize as they were treated to a string of sensational reports dwelling on human tragedy and devastation. Newspaper and radio reporters encountered difficulties of their own in telling the public what was happening. Acting as the mayor's public information officer, Hill restricted reporters from the waterfront area for their own safety but instituted regular briefings to keep them from becoming a nuisance.

Itself a victim of devastation, the local Red Cross chapter could do little initially to discharge its relief duties. According to the report of the Fourth Army disaster team, "No requirements were known by the Red Cross and no requests for assistance could be obtained."[8] Although the mayor stirred up controversy by criticizing the Red Cross in a newspaper interview, Dr. C. Paul Harris, who supervised salvage operations at the docks, probably put the organization's performance in perspective when he stated that volunteers worked well but not paid officials. Around noon on the 16th, when additional blood plasma, donors, and transportation were desperately needed, Dr. Harris was unable to locate anyone from the Red Cross and turned to Army representatives, who obtained blood elsewhere and flew it to Galveston. In the meantime, Dr. Harris organized volunteers to solicit donations up and down Main Street, set up a blood typing center behind city hall, and personally found doctors and nurses to carry on this work.

Regular units of the armed forces were in fact very important to the relief effort during its initial stage. Arriving in Texas City about 5:30 P.M. on 16 April, General Jonathan Wainwright, the Fourth Army commander, immediately noted the absence of coordinated effort. After conferring with the mayor and other officials, he concluded that only the Army had the capability to provide immediate relief on a meaningful scale and assigned Brigadier General J. R. Sheetz to command the Fourth Army's disaster relief team. Mayor Trahan and Governor Buford Jester, who also visited Texas City that afternoon, both concluded that a declaration of martial law was unnecessary, so relief became the sole function of the army team. Together with volunteer organizations, Fourth Army field kitchens fed several thousand persons daily in the business district and delivered vital medical supplies from its own stocks and those of other units. Nearby bases provided temporary quarters for refugees, and personnel opened up and staffed a center at Camp Wallace, a deactivated World War II facility nine miles west of Texas City. The army planned to relinquish control to the Red Cross as soon as a disaster team arrived from national headquarters in St. Louis, but the transfer did not take place until 23 April.

High school gymnasium in use as temporary morgue.
Courtesy of Rosenberg Library, Galveston.

"You Have Violated a Traffic Law"

Having extracted most of the injured by midafternoon, rescue workers began bringing out the dead. Death on this scale was well beyond anyone's imagination or experience. There was no plan or facility suitable for handling hundreds of bodies strewn around the North Slip, many of which were torn apart or mutilated beyond recognition. Again, improvisation was the order of the day. A system of sorts was established when a Galveston undertaker named William Tipton obtained the vacant McGar Garage for use as a mortuary. This was conveniently located across the street from the high school gymnasium, one of the few buildings with sufficient floor space to set out the large number of bodies for identification. Initially, problems arose when some of the dead were taken to funeral

homes in neighboring communities without notification. This prompted the mayor to appoint a three-person dead body commission which had to provide written authorization before bodies could be removed from the town.

The dead were scattered over a wide area around the docks. From the wreckage of Frank's Café at the head of the North Slip, where twenty bodies were found, from Warehouse A, containing thirty, from Monsanto, with more than a hundred, the dead were brought to the garage for identification and cleaning. Both were difficult tasks since most bodies were covered with the same oily, sticky substance as the wounded, and in many cases, the force of the explosion had split open heads, blown off clothes, and severely twisted arms and legs.[9] Fortunately, experts from the Identification Bureau of the Texas Department of Public Safety and the FBI arrived during the day and applied their forensic skills to the most difficult cases. For identification, to the left ankle of each victim was attached a tag from the Port Arthur Police Department which read: "You have violated a traffic law. Port Arthur Police Department." Approximately 150 embalmers, many of whom were students at mortuary training schools in Fort Worth and Houston, worked with whatever equipment was at hand. Sand and tar paper were spread on the floor so embalmers would not slip in the blood and embalming solution. With few preparation tables available, cots, stretchers, counters, and even a grease rack served as substitutes. After embalming, the remains were transferred to the high school gymnasium for identification; as the afternoon progressed, the gym floor began to fill with long rows of bodies wrapped in blankets of Army brown and Navy gray. By supper time on the 16th, there were about two hundred, waiting to be identified and claimed.

The Turning Basin

This disaster was all the more complicated because it encompassed the marine as well as the land environment. Response from the waterside was also sporadic and fragmented and, with eventually horrible results, took place in virtual isolation from efforts on land.

J. G. Tompkins, vice-president in charge of Lykes Brothers' operations at Galveston, had arrived at his office on the morning of the 16th shortly before 9:00 A.M. He remembered that "some of the boys in the hall were looking out of the window and told me that the SS *Grandcamp* was afire."[10] His first call was to Swede Sandberg to inform him that Lykes Brothers officials were on the way and that two tugs would be dispatched in case they were needed. He recalled that Sandberg did not seem particularly concerned. At the time they were summoned, these boats, the *Albatross* and *Propeller*, were docking another ship and could not start the forty-five-minute journey to Texas City for about ten minutes. They had just reached the entrance to the turning basin when the *Grandcamp* exploded. Angelo Amato, captain of the *Albatross*, remembered that fierce fires, thick smoke, and intense heat made it impossible to reach the docks. Under instructions only to assist in fighting the fire on the *Grandcamp*,

Coast Guard fire boat at Monsanto plant. Moore Memorial Public Library, Texas City.

the tugs ignored the *Keene* and *High Flyer*, pushed onto a small island opposite the docks, and launched their lifeboats. After picking up survivors from two small tugs which had been at the south end of the turning basin at the time of the blast, they returned to Galveston.

Next on the scene was a small Coast Guard boat, C-64309, which arrived shortly before 10:00 A.M. During the day, this boat and a number of small craft belonging to the Coast Guard and Army Corps of Engineers picked up casualties from the docks and the turning basin. Coast Guard boats with fire equipment were able to extinguish some fires on the wharves, but the situation remained basically out of control. Of greater significance was the buoy tender *Iris*, commanded by Lieutenant Roy Sumrall. Before leaving base at Pelican Island in Galveston, Lieutenant Sumrall made several telephone calls but could not find out what had happened. Knowing little more than that a ship had blown up, he pressed ahead with nothing more specific in mind than rendering whatever assistance was possible.[11] As had the tug captains, when he arrived about 10:30, he found the turning basin heavily enveloped in smoke and flame. Together with oil and debris, he noticed what he thought were flour sacks until he saw hands and feet attached. Moving along the outer or eastern edge of the turning basin, the *Iris* tied up on the south side of Pier B. Although the *High Flyer* and *Keene* were on the other side of the pier, Lieutenant Sumrall later testified that the remains of a warehouse and heavy smoke substantially obscured these ships, and he was at first unaware of their identities. Dense clouds of smoke, fumes from burning chemicals, and heat sometimes hot enough to blister the skin were a constant hindrance to all boats which approached the docks.

Lieutenant Sumrall dispatched two parties to search for casualties, apparently around Pier B and adjoining warehouses. One of the crew found a doctor and nurse and brought them to the *Iris* to treat the wounded. The crew's mess served as an operating room. "I can still picture crew members standing at the galley entrance," recalled one of these men years later, "dirty, oily, and bloody, eating sandwiches while looking at the injured persons bleeding all over our mess deck."[12] Having accumulated a considerable number of wounded, the *Iris* left for Galveston about 1:00 P.M., when the Coast Guard radio van in Texas City messaged that a large explosion was likely. As soon as the injured and dead were transferred to ambulances, the *Iris* pushed off again for Texas City, arriving about 3:00 P.M.

Testimony and memories conflict with respect to the nature and timing of events on the waterside that day. Crew members from the *Iris* did board and search the *Keene* and *High Flyer* during their second visit. One of them, Carpenter's Mate Leo Falgout, encountered a Roman Catholic priest, Father Adrian Record, on the docks.[13] Incredibly, they discovered about a dozen crewmen still on the *Keene*. After assisting these men off the ship, they crossed over to the *High Flyer* but found no one during a two-hour search. Falgout remembered smoke but no flame coming from the holds when they first boarded the ship, which had its hatches blown off. At the time they completed searching the ship, which was about 6:00 P.M., flames billowing out of the holds created heat so intense that Falgout was unable to look into the holds to see what was happening. On his way back to the *Iris*, he encountered an Army officer who asked if the *Iris* could tow the *Keene* and *High Flyer* out of the slip. He relayed this request to Lieutenant Sumrall when he reached the *Iris* and informed him of the fire on the *High Flyer*. Lieutenant Sumrall later testified that he felt it wasn't safe to attempt towing operations because of heat, gas fumes, and poor visibility. Having received another warning that an explosion was imminent, he backed his boat away from Pier A and returned to Galveston with more wounded and dead.

Lieutenant Sumrall was not the only one who failed to understand the significance of the fire on the *High Flyer*. After searching the ship with Carpenter's Mate Falgout, Father Record sought out the base of search operations at the waterfront. When he reported his information, "I was told by a fireman that they had no equipment with which to operate and by a sailor that he had been asked to go aboard but had declined to do so. I assumed that the sailor involved . . . had decided just to let the boat burn. I had heard vague and contradictory rumors regarding the cargo but did not realize the nature of it until later."[14]

The fact that search and rescue activities from the land and waterside were conducted in mutual ignorance meant that participants failed to share critical information that was of important common concern. Contact between crewmen from the *Iris* and other small boats conducting search and rescue operations along the docks was also intermittent. The only communications between the *Iris* and anyone on land were exchanges of rumors about possible explosions relayed to the *Iris* by the Coast Guard radio van. No evidence exists to indicate that Lieutenant Sumrall, the senior officer afloat, provided city hall with information about conditions from his perspective on the turning basin. Sandberg, who was monitoring search and rescue activities at the docks for the mayor, later testified that it was not until about 8:00 P.M. that he realized the *High Flyer* was still in port. Most important, when Father Record and Carpenter's Mate Falgout reported fire in one of the *High Flyer*'s holds, no one at the docks appreciated its disaster potential or passed this information on to city hall.

5

The *High Flyer*

OVERVIEW

Among the most serious effects of poor communications was
mutual ignorance of threats between those operating on land
and those waterside. As a result, a critical opportunity for remedial action was lost when the *High Flyer* caught fire because
those who might have been able to help were not informed and
did not know that fertilizer was part of the cargo. Sixteen hours
after the *Grandcamp* blew up, the *High Flyer* disintegrated in a
huge explosion, killing and wounding only a few more persons
but adding greatly to property devastation. Although more
explosions and fires occurred at the tank farms throughout the
17th, the worst was over. Concerted attempts to coordinate relief got under way that day, although effective fire suppression
at the docks did not occur until the evening of the 18th.

A Critical Oversight

Returning about 9:00 P.M. from its second trip to Texas City, the *Iris*
tied up at Pier 18 in Galveston. Wanting a better understanding of the situation ashore, Lieutenant Sumrall immediately drove to Texas City with
his second-in-command, Lieutenant (j.g.) John Whitbeck. Arriving about
an hour later, they went in search of the *Keene* and *High Flyer*. Although
a firefighter at the scene told them neither ship was burning, they could
not verify this because of darkness and thick smoke. At this point the two

Coast Guard officers encountered an Army lieutenant named Palmey who asked them if the ships could be moved. Sumrall offered to return with the *Iris*, but when Lieutenant Palmey checked with a superior, he was informed that the ships were about to be removed by tugboats.

The news that someone was trying to tow the *High Flyer* was obviously a complete surprise to Lieutenant Sumrall. Such failures of communication resulted in additional devastation to Texas City that night. Unbeknownst to Lieutenant Sumrall, after a frustrating day seeking information about the condition of the *Keene* and *High Flyer*, Lykes officials had finally succeeded in mounting an operation to remove these ships from the Main Slip. Unfortunately, the effort was so long in coming that it did nothing to prevent further disaster.

James Tompkins, West Gulf Division manager of Lykes Brothers, had concluded his telephone conversation with Swede Sandberg about the fire on the *Grandcamp* shortly after 9:00 A.M. He was dictating a message to headquarters in New Orleans to inform them that he was sending tugs to Texas City when the *Grandcamp*'s explosion shook the building. Unable to reach Sandberg again, Tompkins still hoped to obtain information about the condition of the *High Flyer* and *Keene* either from the two tugs—*Propeller* and *Albatross*—or from company engineers who were to inspect repair work on the *High Flyer*. Just to make sure, he also dispatched two other officials by automobile to find out about the condition of the crews and ships.

Tompkins could easily see pillars of black smoke billowing into the sky from Texas City. But despite the manifest urgency of the situation, information about the crews and the condition of the ships proved elusive. About noon, the two men who had been sent to the docks earlier returned to the Lykes office with no useful information because they had not been permitted to go to the docks. Someone met the tugs when they returned to see if crew members from the *High Flyer* or *Keene* were among those who had been picked up. The captain of the *Albatross* reported that intense heat, fumes, and smoke had prevented an inspection of the ships. A smaller boat was sent during the afternoon, but again heat and dense smoke apparently frustrated efforts to assess the condition of the ships. It was only when injured crew members from the *High Flyer* taken to a

View from Galveston, morning of 16 April. Moore Memorial Public Library, Texas City.

Galveston hospital telephoned the Lykes office that company officials learned the ship had suffered extensive damage and that its mooring lines had parted and it was now lodged against the *Keene*. Ironically, the only Lykes employee able to visit the docks was a company physician, Fred Aves, who called Tompkins about 1:00 P.M. and reported that although he had not boarded either ship, he had been told that both were abandoned.

As indicators of the prevailing confusion, what Lykes officials did not do is as interesting as what they did. Nowhere in his extensive testimony to the Coast Guard investigating board concerning his activities that day did Tompkins mention Captain Petermann, master of the *High Flyer*, although Petermann stated he had asked the chief steward to call the Lykes office with the news that all the crew had left the ship and most were in local hospitals. Moreover, Lykes officials apparently never contacted the

Coast Guard, which might have provided some information about conditions at Pier A and, later, about the fire on the *High Flyer* that Carpenter's Mate Falgout and Father Record had discovered.

Ignorance of the *High Flyer*'s condition became a danger for those on land as well as those at the turning basin. Available evidence indicates the confusion which prevailed among the responders at the time, making it extremely difficult to ascertain the precise sequence of events on the evening of the 16th; as noted earlier, even sworn testimony of major figures conflicts on important points. In the early afternoon, Swede Sandberg was assisting the mayor by monitoring search and rescue efforts at the docks. Almost two years later, he stated in a court deposition that he had gone to the docks twice but that smoke had obscured his view of the Main Slip. He had then assumed that the *High Flyer* was no longer in port.[1] Sandberg stated that he was at city hall when Captain Volney Shawn of the Houston Police Department arrived about 8:00 P.M. and informed him that the rescue workers at the docks were again being evacuated because a ship which contained explosives was on fire. Sandberg recalled that he had refused to believe him at first, but Chief Ladish contradicted this when he testified that he had spent part of the afternoon helping Sandberg ascertain what was in the *High Flyer*'s cargo. Moreover, R. C. Watley, a Houston police officer, remembered that "numerous reports" were received throughout the day that the *High Flyer* would blow up.[2]

Although their concerns were much less focused, townspeople remained in utter dread about threatened explosions. Their fears were fed by citizens cruising through town in automobiles mounted with speakers as well as unconfirmed reports broadcast on commercial radio. All this caused periodic evacuations of considerable numbers of citizens to the northern edge of town, where they waited for more information or simply decided to leave altogether. In the late afternoon, someone with a loudspeaker created the third major alarm of the day by announcing that everyone without a gas mask should leave at once and that an ammunition ship was on fire and was expected to explode any minute. Police officers, national guardsmen, and civilians thereupon raced through the city, warning people to flee for their lives. Those preparing a late meal left food on stoves and tables as they took to cars and bicycles and even left on foot.

Unfortunately, a number of doctors, nurses, and embalmers also departed before Mayor Trahan heard about the rumor and squelched it. This was probably the occasion when, hearing that the mayor had ordered a total evacuation of the town, the Army began preparing Fort Crockett and Ellington Field for an influx of refugees.

Ordinary citizens were not the only ones left in a state of total confusion by the paucity of reliable information. Certainly, the news of a fire on the *High Flyer* was never passed on to Sandberg by either the Coast Guard or the fireman to whom Father Record had spoken. Sandberg later testified that he learned the *High Flyer* was still in port in the early evening only when he noticed workers leaving the dock area in response to a rumor that a ship with ammunition on board was about to explode.[3] Sandberg's court deposition reveals that he was never able to satisfy himself that there was a fire on the *High Flyer* until this point because of numerous other fires and heavy smoke at the docks.[4]

Two critical hours passed after Falgout and Father Record left the *High Flyer* before even those directing search and rescue at the docks understood that its cargo was burning. About 7:30 P.M., Captain Shawn came to city hall and informed Chief Ladish and Sandberg that the rescue workers were pulling back. Sandberg, who gained universal respect among the townspeople for the energy and intrepidity he showed that day, called Tompkins and requested that Lykes try to pull the ship away from the pier so workers could return to the waterfront. "Jimmy," said a very agitated Sandberg, "can you all get the *High Flyer* away from here? There has been an erroneous report—a rumor or a report, that the *High Flyer* has ammunition or explosives aboard and as a result our rescue workers are coming out of that vicinity where there are dead and possibly wounded."[5] Tompkins replied that he had been unable to get tug crews to volunteer for this task. Apparently, Sandberg was now aware that the cargo consisted of sulfur, ammonium nitrate fertilizer, and dismantled boxcars. Seeking more information that might help Tompkins's effort to find volunteers to crew tugs, Sandberg related that he sought expert opinion from chemical engineers about the threat. "They did say this," Sandberg told the Coast Guard board, "that if the fertilizer or ammonium nitrate caught on fire and burned a sufficient length of time that it reached a point of explosion, that it would just have the effect of a brief explosion, and that as long as men were not directly in the ship [*sic*] they saw no danger."[6] Sandberg called Tompkins again with this information, adding that the heat had abated, the slip was clear, and there was no question in his mind that tugs could reach the *High Flyer*. He did say that one anchor of the ship was down and would have to be cut, but he did not mention that the ship was tangled up with the *Wilson B. Keene*.

This was the first indication Tompkins had had all day that any possibility existed of removing either of the Lykes Brothers ships. He now confronted the formidable problem of obtaining volunteer crews for this potentially dangerous operation. He called Bay Towing Company in Galveston and explained that rescue operations would be jeopardized unless the *High Flyer* was removed from the docks. An official there thought tugboat crews would not undertake such dangerous work since the ship was deserted. Nevertheless, Tompkins went to the company office and personally appealed to the tug crews, explaining what was at stake and what was in the holds of the *High Flyer*. "In a very short time somebody said, 'I understand the situation, I will go,' and up went a hand and up went another hand and, finally, before you could almost say 'Jack Robinson,' practically all the hands in the room were up."[7]

At approximately 8:30 P.M., the tugs *Guyton*, *Albatross*, *Clark*, and *Miraflores* cast off for Texas City. On one boat were two Lykes officials and an acetylene torch team assigned to cut the anchor chain. When the tugs arrived in the turning basin about fifty minutes later, *Guyton* leading and *Albatross* close behind, Lykes officials and volunteer crews found conditions much worse than Sandberg had described. Intense heat, heavy smoke, and fumes from burning sulfur that made breathing difficult demanded great resolve on the part of the volunteers and Lykes officials. According to Angelo Amato, skipper of the *Albatross*, the fire on the *High Flyer* "sounded like a volcano the way it was burning and all the sparks that was coming out of there. It was a terrible noise."[8] A jumbled mass of logs and piling as well as the manner in which the *Keene* and *High Flyer* were lodged together prevented tugs from moving into the slip. First, the *Guyton* hooked a line onto the anchor chain at the bow of the *High Flyer*, but the line soon parted. Several men, including the Lykes officials, clambered aboard the ship, and the acetylene team on the *Albatross* cut the anchor chain above the waterline. With a new line attached to the anchor chain, the *Albatross* and *Guyton* pulled in tandem. Even after the boarding party moved to the bow and cut the lines entangling the ship with the *Keene*, twenty minutes of effort moved the ship only about fifty feet.

The first hour of 17 April had now elapsed. Ominously, balls of fire began shooting out of the ship's holds. It was time to leave. One member of the boarding party signaled the tugs to cut the towline and pick them up. The *Guyton* held against the ship while the boarders slid down a line onto the deck and then retreated 200 yards into the turning basin. The tug had turned around and was slowly starting back toward the *High Flyer* with a light trained on it when the ship exploded with a tremendous roar. One of the *Guyton*'s passengers remembered seeing a red flash as he was knocked down. Why the tug was not obliterated is a mystery, but the force of the explosion thrust it backward about a hundred yards farther from the pier, disabling it and throwing the crew into severe shock. The *Albatross*, which was just around the bend of the Texas City channel, returned and removed the wounded, while the *Miraflores* towed the *Guyton* back to Galveston. The Coast Guard board of investigation eventually concluded that the ammonium nitrate fertilizer, which was in Hold 3, exploded after becoming contaminated with the sulfur that was stowed in Holds 2 and 4. But like the *Grandcamp*, the *High Flyer* had a large tank of Bunker C fuel oil between Holds 3 and 4 and more fuel in tanks between the bottom of the holds and the keel. It is possible that burning sulfur set the ammonium nitrate on fire, and, similar to what may have happened on the *Grandcamp*, the intense heat melted the bulkhead between the fuel oil tanks and Hold 3, allowing mixture of the two substances and triggering the explosion.

Sandberg's belated recognition about 8:00 P.M. on 16 April that the *High Flyer* represented a threat was a step in the right direction. But the absence of organization and the ensuing confusion caused him to work at cross-purposes to the mayor. By early evening, sulfur fumes from the burning ship were so intense that everyone working around the Main Slip was obliged to wear a gas mask. Patrolman R. C. Watley of the Houston Police Department remembered that the *High Flyer* was erupting fire and smoke "with increased intensity for about an hour before the explosion."[9] About the time the tugs arrived, Sandberg came to the docks and informed some of those working there that he had just learned that ammonium nitrate fertilizer was in the *High Flyer*, and he cautioned them to remain at a safe distance from the ship. Two engineers from the U.S. Bureau of Mines who had arrived on the scene were much more concerned than Sandberg: "By this time I became a little uneasy and nervous," wrote one of these officials, "having found out that one hold of the S.S. *High Flyer* was loaded with nitrate and that the sulfur was burning in an adjacent hold."[10] Although the sides of the ship did not appear to be hot, he did notice a considerable amount of white smoke coming from the decks. Realizing that another serious explosion was possible, the engineers immediately left the docks, advising others to do so as well and initiating a withdrawal of rescue workers toward the center of town about 11:30 P.M. No one had bothered to inform the mayor of this threat, for about 1:00 A.M. on 17 April he broadcast an appeal on the radio for workers to remain at their jobs, stating that there was no danger of further explosions. Additional evidence of poor communication and resulting confusion is apparent in Chief Ladish's testimony that he was advised at 11:00 P.M. by a Houston police inspector that "the situation at the docks was bad" because the *High Flyer* was on fire and all personnel should be pulled out of the area.[11]

After leaving the waterfront, the two engineers returned to their car. They were driving out of Texas City on the highway toward Houston, discussing the possibility of the *High Flyer* exploding, when exactly this happened. As the *Grandcamp* had done before, the power of the blast simply disintegrated the *High Flyer*. A huge column of flame billowed several thousand feet into the air, lighting up the night sky. Fiery balls peeled off from the main mass, and pieces of red-hot shrapnel arched out from the top of the column in a giant fireworks display. When the shock wave reached the two men in their car the motor raced, but the car, momentarily lifted off the ground, did not move. Soon thereafter, a heavy dust or mist—they could not tell exactly which because of the darkness—descended upon the area. Others much farther away were also affected; in Galveston, Bill Simpson, a Red Cross worker who had gone home after a hard day's work, was bounced out of bed by the blast.

Casualties were light this time because most persons had left the dock area. But two of those who did remain were killed and some twenty-four injured. One of the former was a man standing 500 feet from the *High Flyer* who was almost decapitated by a flying piece of steel. This may have been Father William Roach, of St. Mary's Roman Catholic Church, who had labored hard and long to help the injured. The other was Charles Kelly, a student mortician who had come to the docks from the makeshift morgue in the McGar Garage to collect the dead. A Houston Police Department lieutenant suffered severe shock and a broken arm when shrapnel raining out of the sky caved in the top of the vehicle in which he was sitting. Others performing errands of mercy also suffered; a woman Red Cross canteen worker was brutally slammed to the ground by the force of the blast, and shrapnel took off the right foot of a Salvation Army volunteer who had just arrived at the docks with food and supplies. An innocent bystander almost a mile from the docks had an almost unbelievable experience. Mrs. E. A. Anderson had been standing in front of a jewelry store when the *Grandcamp*'s explosion shattered the plate glass window, cutting her across the back. After obtaining medical treatment and resting at a relative's home all day, she ventured out on the street that night. At precisely the moment the *High Flyer* blew up, Mrs. Anderson was in front of the same jewelry store, where the plate glass window had been replaced, and received more cuts on her back when this glass shattered.[12]

Those present at both ship explosions believed that the *High Flyer* blew up with even greater force than the *Grandcamp*. With the exception of a part of the ship that landed on a warehouse, the *High Flyer* also disintegrated into thousands of small pieces of shrapnel, some of which were thrown as far as 6,000 feet. The blast effects completed the devastation of waterfront facilities and started new fires among petroleum storage tanks in nearby farms, particularly at the Humble facility.

Recall that the force of the *Grandcamp*'s explosion had torn the *High Flyer* from its moorings, allowing the ship to drift across the Main Slip and lodge against the *Keene*. When the *High Flyer* blew up, the aft part of the *Keene* was obliterated, leaving the fore part of the ship half sunk in the water, a total wreck. A subsequent survey by the Army Corps of Engineers revealed that the depth in the slip immediately underneath the amidships where the *High Flyer* had been was 65 feet, rather than the normal 35 feet. A 35-ton piece of the ship's hull was later found on the floor of the turning basin.

Those who were outside at the time remembered another extended period when fragments of the ship rained out of the sky. Red-hot shrapnel peppered refining and storage facilities, starting new fires and rekindling some which had been extinguished with considerable effort during the afternoon. Four crude oil storage tanks at the Humble tank farm burst into flame. Three more caught fire later and burned to the ground because water was not available. Some tank fires could not be extinguished because pipe fittings on foam tanks built to drain through applicators were damaged by shrapnel. Shrapnel ruptured a 2,500-barrel tank at the Stone Oil Company refinery, igniting the product and blowing the dome roof about 350 feet. Some 6,000 feet from the docks, a crude oil tank at Republic's west tank farm with only about five feet of oil in it also burst into flame. Adding to the conflagration was an aluminum tank containing isopropyl acetate and a steel tank at the Carbide and Carbon Chemical Company terminal which eventually melted to the ground.

The waterfront was now thoroughly devastated. One of the engineers from the U.S. Bureau of Mines who had been at the docks during the evening of the 16th was "much amazed at the additional destruction" when he returned the next morning.[13] The warehouses on Piers A and B, reinforced concrete structures which had withstood the effects of the previous explosion reasonably well, were now piles of rubble. The last third of Pier A toward the turning basin was obliterated. Concussion stripped off the iron exteriors of the three remaining bulk goods warehouses on the waterfront, C, D, and E, located to the south of Pier B, and twisted their frames into masses of crumpled steel. Most of the contents of these buildings as well as flour sacks in freight cars beside Warehouse C burst into flames. The concrete grain elevator located at the head of the Main Slip was pierced by shrapnel, causing some of its contents to spill out and catch fire. In the morning, an explosion of grain dust finished off this structure. A forty-foot-high conveyor belt which carried grain from ships to the elevator collapsed across the shell-topped road, complicating efforts

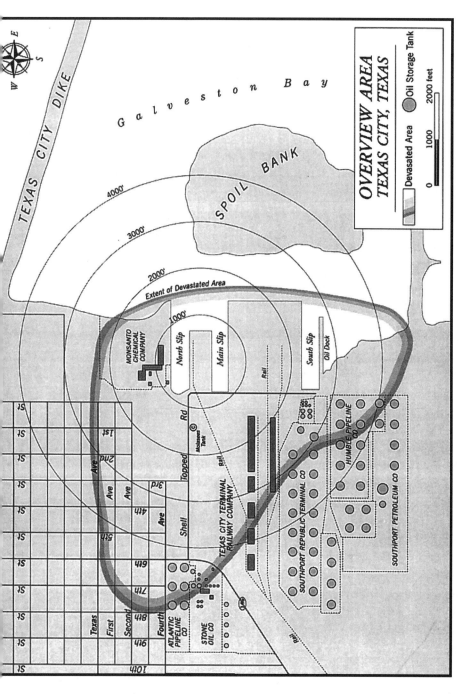

Map 4

to clear the only road to the docks. Fires broke out at the South Slip when blast overpressure and shrapnel twisted and punctured pipes leading from the Pan American refinery to its loading dock.

The area beyond the docks fared little better. Concussion and shrapnel from both ship explosions added to the junkyard of employee automobiles and railway rolling stock, augmented by boxcar wheels and gondola sides, residue of the dismantled railway boxcars aboard the *High Flyer*. Located 1,500 feet from the Main Slip, the Terminal Railway power house and shop building were now totally wrecked. Eight of nine storage tanks at the Carbide and Carbon Chemical Company, opposite the South Slip, were a total loss, as were six other less substantial warehouses directly west of the Main Slip at distances between 1,000 and 4,000 feet.[14] Similarly, houses already weakened by the *Grandcamp* explosion, particularly in the black and Hispanic areas located nearest the docks, had their roofs caved in or their walls collapsed by this second massive concussion. About one-third of the residences of Texas City were now unfit for human habitation.

As extensive as the damage was, it could have been much worse. Many tanks punctured by shrapnel or crumpled by blast overpressure did not explode because they were relatively full and little room was available for expansion and ignition of their contents. The Republic Oil refinery was probably spared highly destructive fires when shrapnel narrowly missed a high-pressure fuel gas manifold and a thermal cracking unit.

Within minutes of the explosion, rescuers rushed back to the flaming waterfront and began extracting those hurt by the explosion. Soon, the sound of wailing sirens cut through the night as ambulances conveyed the injured to a makeshift hospital at the city auditorium. This task appears to have been completed relatively quickly, for the Texas City police radio log indicates that offers of additional ambulances were being turned down within thirty minutes of the explosion. Army doctors and a collecting company of the Thirty-second Medical Battalion which rushed in from Fort Crockett were informed that their services were not needed because medical capabilities were sufficient to handle the additional casualties. Since some of the air force searchlights in use at the docks were destroyed and nothing could be done about the fires, search and rescue efforts were suspended until morning. Not that the remainder of the night was quiet;

about 2:00 A.M., a storage tank at the Humble tank farm blew up with a tremendous roar, and another exploded an hour later.

The dawn of 17 April found a "solemn stillness" hanging over the wrecked buildings, glass-strewn streets, dazed and exhausted survivors, and equally tired rescue workers.[15] Although it was to shift during the day, the northwest wind still held in the early morning, pushing huge clouds of heavy black smoke issuing from burning crude oil storage tanks down the length of the waterfront and out over Galveston Bay, away from town. "One looked always on the silent black billows unfolding endlessly in the sky," wrote a reporter from the *New York Herald Tribune*, "a dark symbol of the tragedy that has crushed Texas City."[16] Later, when the wind shifted, airplane pilots flying to Houston and Galveston from St. Louis with sup-

Cleaning up debris, 17 April. AP Wide World Photos.

plies reported that the smoke extended to the southern border of Missouri. The waterfront was obscured by fierce fires still burning at Monsanto and the warehouses. Daisy Vincent, then a schoolteacher, remembers a stench permeating the town that lasted for days.

A Gloomy Awning of Darkness

Although it may not have seemed so at the time, the worst had passed. Still, the Army Corps of Engineers found the situation on Thursday, 17 April, to be one of "considerable confusion." Because virtually nothing was in place for recovery, several days elapsed before a measure of overall control was achieved. Public fear, reinforced by occasional explosions at tank farms or along the waterfront, remained palpable and immediate. Citizens still retreated to the northern edge of town every time well-meaning individuals in sound trucks broadcast rumors of impending explosions or toxic gas releases. Chief Ladish again tried to halt this practice, and even though some of the press or public never realized it, John Hill began providing regular briefings to the mass media. One study of the disaster identified radio broadcasts as a major source of rumors: "As soon as power was restored to Texas City and the people there could listen to their radios, they began to hear highly dramatic and often alarming accounts of what was going on within a few thousand feet of them. Rumors which were at the very moment being denied by officials in Texas City were, at the same time, being broadcast to the residents over national networks as 'news.'"[17] In the anxiety of the moment, even casual remarks set off panic; when a mortician working at McGar Garage asked for a mask and someone yelled across the room for a gas mask, a report quickly swept through town that poison gas was drifting out of the industrial area.

Recovery efforts finally began to take form during the morning of the 17th. At 10:00, Mayor Trahan convened a meeting attended by his makeshift staff and several other officials on the scene, including Brigadier General Sheetz, commanding the Fourth Army's disaster team, Maurice Reddy from national Red Cross headquarters, and local industry representatives.

Reddy, who had arrived during the night with twelve officials from St. Louis, faced a challenge much like the mayor's—pulling together an organization to integrate efforts of local chapters in southeast Texas and coordinating these with delivery of services provided by his national headquarters. His task was facilitated by Governor Jester's proclamation of a state of emergency that designated the Red Cross as the coordinator of disaster relief activities in the Texas City area and as the recipient for funds and other forms of assistance. The governor's action also assigned the Department of Public Safety responsibility for "all police and rescue activities in the affected area," confirming access control measures already taken by law enforcement officers the day before. Beyond the positive effects of discouraging looting, this facilitated the exclusion of the merely curious, particularly those who would have shown up during the weekend despite appeals by the mayor to stay away. Confirming an action that the mayor had taken on his own, the governor also formally ordered the Forty-ninth Battalion of the Texas State Guard to active duty. Contingents of police from San Antonio, Dallas, and other distant localities arrived early on the 17th to relieve law enforcement volunteers from Houston, Galveston, and other neighboring jurisdictions who had manned roadblocks and patrolled the streets since midday on the 16th.

Efforts to organize fire suppression did not meet with anything like as much success. About two hundred volunteer firefighters, mainly refinery teams and oil fire specialists, gathered at the high school football field on the morning of 17 April for assignment by Louis Shapiro of Pan American, whom the mayor had appointed as coordinator of fire-fighting operations. Theirs was a formidable task, for twenty tanks were burning, six at Humble, and nine posed serious threats. Those at the Richardson refinery had been mostly empty, and they exploded like paper shells when falling shrapnel from the *High Flyer* pierced them, setting off their vapors. Republic Oil had lost three 80,000-barrel tanks and Stone several smaller ones. Two benzol tanks at Monsanto, ignited by the *Grandcamp* explosion, blazed as fiercely as ever.

The scene at the football field exemplifies the difficulties of organizing people who have never before worked together to accomplish a complex, dangerous task. "After a considerable waiting period," teams supplied with

Burning fuel oil tanks near docks. Moore Memorial Public Library,
Texas City.

foamite, foamite towers, and generators moved off to their assigned areas
in order to prevent fires from spreading as much as to extinguish them.[18]
Foamite, the most effective suppressant for tank fires, could not always be
used because either pipe fittings for applicators were damaged or water
was unavailable. Although some tanks burned themselves out during the
morning, two more large crude oil tanks caught fire at Humble, sending
flames shooting high into the air. Fortunately, tanks were spaced well
apart, and synergistic reactions that spread fire and explosions from one
tank to another were rare.

At the waterfront, search and rescue continued during the 17th. A force
about two hundred strong under the direction of C. Paul Harris, a Houston dentist whose family had been in the construction business, system-
atically searched the debris in a 400-yard-wide area to the south of the

grain elevator. Fires and heat, acrid smoke, building rubble, large pieces of ship shrapnel, and wrecked automobiles littering the area made the going difficult. Returning with the *Iris* at 7:30 A.M., an undoubtedly weary Lieutenant Sumrall radioed the Coast Guard truck at city hall that fires along the waterfront were worse than the previous day. As workers equipped with acetylene torches cut shrapnel into manageable sizes, Army and civilian bulldozers, cranes, dump trucks, and other heavy duty equipment removed or pushed aside wrecked cars and rubble. This activity continued throughout the day, although operations were temporarily suspended shortly after 4:00 P.M. when another tank at the nearby Humble tank farm exploded with a roar, spreading flames throughout that property. Canteens operated by the Salvation Army and Red Cross, as well as the Army field kitchen, continued to feed volunteers. By evening, cots and blankets littered the front lawn of city hall, where exhausted firefighters and relief workers rested briefly before returning to their work.

The grim task of collecting the remains of the dead also continued. After working with an acetylene torch to free injured persons and corpses for two days, former marine Art Lee found himself unable to keep down any food.[19] Many victims were half-buried in the oily muck, with arms, legs, and even heads torn away by blast effect and shrapnel. Guided by state and federal identification experts, workers learned to seek personal belongings useful for identifying blackened and mangled bodies. Some hit upon the practice of locating wallets whenever possible and placing these in the mouths of victims. Embalmers worked harder that morning as fifty more bodies were found and brought to the McGar Garage. Morticians found conditions trying: "The garage morgue had a cement floor, and in embalming the bodies there was no way for us to drain into containers; consequently we had to drain the blood on the floor. . . . Embalming fluid was spilled and it drained from the mutilated parts of the bodies. After several bodies were embalmed the floor was an inch or two deep in slush."[20] Many bodies, swollen to several times their normal size, were picked out of the water by Coast Guard and Army Corps of Engineers personnel. Others had still to be located; only one hundred of almost five hundred persons believed to have been at Monsanto were accounted for on the 17th. No one as yet had an accurate idea of how many had perished—it was believed that about two hundred bodies were still in the ruins.

After embalming, corpses were taken to the high school gymnasium. There they were laid in long, neat rows on the floor, wrapped in military blankets for viewing by relatives seeking missing kinfolk. Few of those who found their loved ones there showed great emotion; most were too stunned by what they had already experienced. By Thursday evening, some two hundred names of the identifiable dead were chalked on a blackboard outside the gym. Given the violence of the explosion, many could not be identified right away, and over the next two days the number of unidentified bodies rose to almost one hundred. Until the Texas Department of Public Safety achieved a measure of control over casualty listing, multiple "death lists" with errors gave relatives unnecessary anguish or false hope. A few undertakers caused further confusion by taking bodies to mortuaries in other towns without notifying anyone. Victor Landig, a local undertaker who chaired the mayor's dead body commission, issued an appeal to all out-of-town funeral homes to return bodies for inspection by identification experts. Among those who learned the fate of her loved one during the day was Mrs. Dallas White. She located her husband at John Sealy Hospital in Galveston the day after the *Grandcamp* exploded: "At last my search has ended," she wrote. "Bill and I found his father, my husband, just before he died at 12:50. It was some comfort to tell him good-bye."[21]

Unidentified bodies were only part of a much larger problem of locating missing persons, for like most disasters, this one disrupted normal living patterns and scattered people across the surrounding countryside. Slightly more than half of the work force resided in neighboring towns, and an estimated 40 percent of the population of Texas City fled for varying periods of time. Most residents who left went to friends or relatives until told they could return. Some were taken in by good Samaritans. About seven hundred of the least fortunate found shelter at military bases. As noted earlier, because a comprehensive system was not instituted, spontaneous, uncoordinated dispositions of victims made it especially difficult to locate individuals. The fact that a reliable, comprehensive system for locating missing persons never emerged is attributable to communications and organizational problems rooted in the absence of preparations for this sort of event. Matters were further complicated because, despite restoration of severed trunk lines within a few hours of the *Grandcamp*'s explosion, most residential telephones were inoperative, making personal communication in and out of Texas City almost impossible.

The Red Cross assumed most of the burden of responding to requests for information about residents. Prompted by a steady stream of gruesome radio reports and newspaper articles that riveted the attention of the entire nation, inquiries about loved ones poured in from around the United States as well as from several foreign countries. In all, some 27,000 requests were received. The Galveston Red Cross chapter set up an inquiry staff of 250 persons organized into six-hour shifts of fifty persons each which operated on a twenty-four-hour basis for about a week. The chapter also supplied workers for another office at Texas City. In addition to the prevailing confusion and uncertainty of the moment, the search for information was hampered by the lack of a city directory. Boy Scouts and church youth groups were drafted to try to locate those who might have remained in town. Without systematic record keeping, omissions and duplications compounded the anguish of uncertainty. Several days passed before the Red Cross was fully mobilized and assumed control of relief operations.

Sporadic communications and absence of control continued to hamper relief efforts throughout 17 April. Several mistakes occurred because the Army transmitter sent to Galveston failed, preventing a direct radio link between the base of operations at Fort Crockett, Texas City, and the Fourth Army until the next day. Because almost two hours elapsed before the Coast Guard radio truck relayed a message that army personnel were not needed, ten medical officers and a collecting company from the army units at Fort Crockett were sent to Texas City after the *High Flyer* exploded.[22] On the evening of the 17th, the disaster team made extensive but unnecessary preparations to ready Fort Crockett for an influx of refugees because General Sheetz could not confirm a report that the mayor had ordered an evacuation of the town. Poor communications may have been responsible for some groups arriving at Texas City without being summoned; at about 10:00 P.M. on the 17th, a composite unit of former guardsmen, veterans, and army reservists arrived from Beaumont. As the Texas State Guard *Journal* for the day succinctly states: "Since this unit had not been ordered into the area and was not self-sustaining, they were fed by the 49th Battalion and were sent back to Beaumont."[23] While those who had worked almost nonstop for thirty-six hours did need relief that evening, most of the volunteers and many supplies were no longer required. Nonetheless, this was never effectively communicated to the outside world, for commercial and amateur radio broadcasts continued to appeal for assistance.

On Friday morning, 18 April, fires at the Humble and Republic Oil tank farms still emitted thick, black smoke which, joined by the residue of burning benzol at Monsanto, rose 3,000 feet in the air, forming what a Monsanto official called "a gloomy awning of darkness." As the day wore on, improving organization and hard work brought progress. Glass was swept from streets, new windows were installed, the spilled contents of shops were picked up, and merchants resumed business where possible. Telephone, electric, and natural gas services were restored in homes,

Body recovered from the turning basin is hoisted onto pier. Moore Memorial Public Library, Texas City.

shops, and offices sufficiently sound to allow occupancy. The frantic pace of activity at the auditorium and clinics abated as the focus of medical care shifted to crowded hospitals in Galveston and nearby communities, where tired doctors and nurses still struggled to treat wounds and contain the trauma of hundreds of seriously injured persons. Most of the fifty victims who died in a hospital did so within a week; about 120 others were discharged. On Friday, John Hill was able to tell the press "we're in good shape now" insofar as threats of catastrophic explosions were concerned. Even so, much remained to be accomplished. Hill broadcast an appeal for citizens who had other places to stay not to return yet and for sightseers not to visit Texas City. Roadblocks were maintained to ensure that no unnecessary visitors showed up.

Hindered by smoldering fires in ruined warehouses and building rubble, the search for bodies around the waterfront continued at a brisk pace on the 18th. Workers were able to enter much of the Monsanto property despite a large plume of smoke rising from a burning benzol tank. Wrecker trucks began removing damaged automobiles in the parking lot just west of the plant. The entry of the Fourth Army disaster team journal for 19 April states: "The waterfront area . . . is still in a shambles; cleanup and repair in the area are progressing rapidly. The dead are still being removed from buildings and areas which have been previously inaccessible. A tremendous job remains."[24]

Despite a thick layer of oil and debris on the surface of the turning basin, searchers working from boats found more bloated, mangled bodies. Seven were picked up around the docks on Friday, twenty-two on Sunday the 20th, and nine more as late as the 25th. Others were discovered along the dike that jutted into Galveston Bay and as far away as the Galveston-Freeport entrance to the intracoastal canal. The largest concentration was believed to be at Monsanto. On Saturday, company officials knew that of 449 employees or contract laborers believed to be at the plant when the *Grandcamp* exploded, 90 escaped unhurt, 43 were known dead and were identified, 79 were missing, 115 were hospitalized, and 122 were "believed safe."[25] Considerable time passed before definite information fell into place: the figure of 295 persons listed as missing on Monday was eventually cut by two-thirds, but on Sunday, as the death toll approached 400, the dimension of human loss became tragically apparent.

The sheer number of dead created a crisis of sorts on Sunday. By this time, bodies or body fragments had lain in the open or in debris so long that embalming was almost impossible. Either these had to be buried almost immediately or a cold storage site large enough to hold 150 corpses had to be located in order to preserve them for further identification. On Tuesday the 22nd, these bodies were moved to cold storage vaults at Camp Wallace, a vacated military base ten miles away. By the 25th, two hundred persons were still listed as missing, and almost one hundred unidentified bodies were in the storage vaults.

Suppressing large hydrocarbon fires is never easy. On the 18th, fires were still burning in storage tanks and along the waterfront. A large explosion occurred about 2:00 A.M. and several minor ones during the day. On Saturday, an intense crude oil fire at the Humble tank farm caused a spectacular "boil over" into an adjoining tank. Information about fire fighting is quite sketchy, but it appears that efforts at the docks and at tank farms were pursued independently of one another. Refinery fire-fighting teams succeeded in extinguishing fires at Republic Oil, but many others were allowed to burn themselves out. If a staff report of the *National Fire Protection Quarterly* is correct, concerted fire suppression at the docks did not begin until the single shell-topped road to the waterfront was finally cleared of debris, allowing pumpers from the Houston and Pasadena Fire Departments under the direction of Houston's Chief Whittlesey to draw water from the turning basin.[26] Even with the help of Coast Guard fire boats, waterfront fires smoldered until Monday. That day the last of the tank farm fires was extinguished, while the last fire at Monsanto burned itself out on Tuesday, the 22nd.

The daily situation report of the Fourth Army disaster team for Saturday, 19 April, states that the emergency was gradually subsiding and organization on the part of the civil authorities was beginning to take hold.[27] But grieving, anxious citizens, traumatized by the events of the past several days, remained as apprehensive of further trouble as ever. Even the slightest rumor of serious trouble precipitated panic. On Sunday, someone detected the odor of naphtha from leaking tanks at a Republic Oil facility on the edge of the industrial area. When the mayor imposed a no-smoking order and guards were posted near the facility, reports of an impending explosion prompted hundreds of residents to flee the town. The fact that commercial radio continued to broadcast rumors irritated a very tired John Hill. Another scare occurred on Tuesday morning, when ammonium nitrate fertilizer in the ruins of Warehouse O next to the North Slip caught fire. Rising orange smoke quickly generated rumors that a new explosion was imminent. The mayor immediately went on the radio and promised he would warn everyone if there was a real chance of explosion. After about an hour, the smoke died away, and a Coast Guard fire boat moved in and wetted down the area. Later a barge with two high-pressure hose trucks washed the debris from the docks into the water. Still, anxiety remained; townspeople were severely frightened several weeks later when a tanker at the South Slip caught fire and burned for a while. "One cannot go through these experiences with death all about and be the same ever," wrote George Reck, a Lutheran minister from Houston. "Even the children seem to be trembling inside. No one [at the Lutheran Mission] was seriously injured, although healing wounds about the face reveal how dangerously close death was."[28]

The first attempt at collective healing occurred on Saturday, 19 April. That afternoon, even as bodies were still being pulled from dockside rubble, about fifteen hundred informally dressed, dazed, and grieving people gathered at the high school football field against a backdrop of smoke from burning oil tanks for a forty-five-minute memorial service organized by the ministerial alliance. After cleaning up their own damaged buildings, churches began holding memorial services and funerals. On Sunday, thirty-two processions made the eight-mile trip to the cemetery. "We are worn out with work," said John Clay, a sexton. "We dig all day, and then when it gets dark we light our lanterns and keep on digging. It's sad work, but we are fixing a place for all those people blowed away to their Lord."[29]

6

Aftermath

OVERVIEW

As the last of the wounded were treated, the dead collected, and fires extinguished, stunned citizens and weary relief workers took stock of human and property losses. With outside assistance, local leadership and business initiative quickly restored vital services and began the tasks of rebuilding and even expanding municipal infrastructure and industrial facilities. Although the scope of investigation guaranteed that many questions of culpability would be ignored, the nature of the experience did prompt some changes toward better preparedness for disasters.

Even if the town's citizens were not yet convinced of it, by Tuesday, 22 April, the danger had passed. The pall of black smoke which had hovered over the town dissipated as the remaining crude oil tank fires were extinguished or burned out. The day before, most of the several hundred volunteers who comprised the backbone of search and rescue efforts had departed, and military units began returning to their bases. On Thursday, John Hill broadcast an appeal for those who had fled to come back if their houses were habitable. Nonetheless, severe trauma and physical disruption lingered long after the immediate crisis had passed. Several months elapsed before retail business returned to normal and damaged houses were restored. A few families remained on Red Cross financial assistance into 1948. The last of the rubble was not cleared from the docks for over

Mourners at funeral service held in June for unidentified victims; Rev. Frank Johnson in background. Moore Memorial Public Library, Texas City.

a year, and the wreckage of the *Keene* sat in the Main Slip until mid-1949. Nine years went by before the Terminal Railway and its insurers concluded their legal dispute over liability for damages. Not until 1957 did relatives of those killed in the ship blasts receive monetary compensation from Congress for the death of their loved ones, following rejection of their claims in federal court. For 844 widows, widowers, dependent children, and elderly parents, the loss was permanent, while hundreds of the injured were condemned to live with physical afflictions for the rest of their lives. Fifty years later, a residue of sorrow remains among survivors, relatives, and friends that will endure until that generation and their children pass completely away.

Relief Begins

The last body was recovered from the dock area on 11 May. Although it proved impossible to determine exactly how many people had been killed, confirmed losses were stunning. After careful investigation, the Red Cross and the Texas Department of Public Safety counted 405 identified dead and 63 unidentified dead. Another 113 persons had simply disappeared without a trace and were listed as "believed missing," bringing the total to 581. Of these 27 were women, 20 were sixteen years of age or under, and 3 were younger than six.[1] Estimates of wounded are even more problematic since some of the injured fled town immediately and were treated elsewhere. When the unidentified dead were buried together at a memorial cemetery on 22 June, 380 of the approximately 3,500 persons who had sustained injuries were still hospitalized. As noted earlier, although not all casualties were actually residents of Texas City, the aggregate total of almost 4,100 persons killed or injured was equivalent to about one-quarter of the town's estimated population. These figures do not include obstetrical cases, depression, or other forms of psychological trauma. Mental breakdowns were rare and suicides nonexistent, in part because everyone shared in the trauma and human misery was so widespread. Some, like Rosa Lee Curry Eelbeck, woke up screaming from nightmares for weeks afterward. A feeling of numbness which lasted for months came over other survivors. And there were other "casualties" as well. Twenty years later, a funeral home director named Fred Linton recalled, "The deaths weren't confined to those days. We had more funerals in August, September, and on into the fall than we had ever had. People were dying because they had taken all they could stand. They burnt themselves out working in the disaster and they just died."[2]

While survivors were thankful to be alive, many faced months of anxiety and inconvenience before their lives returned to more or less normal paths. The 40 percent of the population that fled Texas City in response to warnings or rumors represented the largest, albeit a temporary, problem. In a town with an already-severe housing shortage caused by rapid growth, about 2,000 persons were left temporarily homeless because 539 dwellings were condemned as unsafe, compelling these people to stay with family or friends or at Camp Wallace. Because some 1,100 automobiles had been wrecked by the blasts, transportation was a temporary hardship for workers. Schools and municipal property, especially city hall, had sustained an estimated $1 million in damages and required extensive repairs.

Sanitation presented critical problems in the immediate aftermath of the disaster. Galveston County's public health authority initiated emergency operations in Texas City on the 17th and rapidly extended its activities with assistance from the state and other local governments. Food left on tables by fleeing citizens and in refrigerators without electricity spoiled quickly and, along with uncollected garbage, posed a potential sanitation problem. Labor gangs under police and Health Department supervision were organized to search vacant houses to remove food, and the mayor issued a proclamation requiring all groceries and restaurants to obtain clearance from the Health Department before resuming business. DDT was sprayed throughout the town in order to prevent proliferation of rats and flies. Public health personnel also assisted in the laborious task of locating displaced and injured persons; community leaders in neighboring towns were contacted, and two houses in each block were surveyed to obtain information about displaced persons. In Texas City itself, the town was divided into sections, and public health nurses visited every occupied residence. As noted earlier, utilities had survived the blast effects of the ship explosions with little damage. Natural gas lines sustained only one break outside the dock area, and the system was fully operational by the afternoon of the 16th. Except for the waterfront, the Community Public Service Company restored electric power throughout most of the town with the assistance of emergency crews from Houston Lighting and Power Company.

As a measure of calm and order returned to Texas City, municipal authorities turned their attention to repairing widespread damage to stores and residences. In cooperation with the Health Department, John Hill organized volunteer building inspectors from southeast Texas to survey 120 commercial buildings and 3,500 houses. By Tuesday, city authorities knew that 10 percent of the former were unsafe: about 130 had suffered damage, particularly to their interiors. Some residential areas were in worse condition: 90 percent of all homes in the town sustained at least minor damage, but inspectors discovered that over five hundred houses, particularly the simple frame structures in black and Hispanic neighborhoods nearest the docks, needed major repairs or rebuilding before they could be inhabited. Volunteer carpenters from Houston and other communities came to Texas City during the weekend of the 19th and 20th and made essential repairs to houses which were sound enough to be reoccupied. In six months' time, about three-quarters of all those that could be repaired were fixed. Spurred by continued population growth, a year after the disaster, five hundred new houses had been built, and almost nine hundred new residences or apartments were under construction. Many of these, particularly an entirely new section called Chelsea Manor, were of a temporary type used in military construction. Much of this activity was supervised by a hastily created Housing Authority initially chaired by Ray Mariana of Carbide and Carbon Chemical Company, an agency which still exists today.

The disaster served as the occasion for additional improvements in municipal infrastructure. Possibly to avoid difficulties citizens might encounter in obtaining new insurance, the city enacted construction ordinances which met the requirements of the Southern Building Code. In rebuilding the Fire Department, more full-time personnel were hired, the single fire station was enlarged, and two others were built, one in the Heights area and the other at the south end of town. The next year, voters passed a bond issue to pave more streets and put in sanitary storm sewers.

As noted earlier, most of the town's commercial establishments were in disrepair. One hundred thirty retail stores and their stocks sustained an estimated $1 million in damages. The docks and their facilities were virtually obliterated. Besides the three ships, valued at $1 million for insurance purposes, the blasts and fires destroyed the offices and warehouses of the Terminal Railway, most of its locomotives, and 350 railcars belonging to various lines. Located directly across the North Slip and receiving the full force of the *Grandcamp*'s explosion, the Monsanto plant represented a $20 million loss. Because they were more distant from the docks, refineries received only minor damage, although half of the 250 storage tanks in the 25,000 to 80,000 barrel range sustained some damage, and 22 percent were totally destroyed.[3] Eight of the latter were at the Humble Oil tank farm, about 3,000 feet from the Main Slip. Including the three ships, total property losses approached $100 million, equivalent to $700 million in monetary value for the mid-1990s.

While the larger corporations had sufficient resources to rebuild and even expand their facilities, individuals and small businesses had little to fall back upon except insurance policies. At the time, Congress occasionally appropriated funds for flood relief, but only on a case-by-case basis. The Texas Legislature did rebate municipal and school taxes for three years, but no equivalent of today's Federal Emergency Management Agency existed to rush personnel and resources to the scene and provide emergency housing and grants for individuals and businesses under the aegis of a presidential disaster declaration. Rather, relief and recovery activities went forward in a decentralized manner; volunteer agencies, churches, companies, individuals, and, to a lesser extent, municipal, county, and state governments assumed or shared most of the burden. In most respects, Texas City was still a small community, and many problems were dealt with informally but effectively. Almost fifty years afterward, Curtis Trahan still remembers how someone of ability and integrity in the community always stepped forward and took responsibility for solving problems that arose in the aftermath of the disaster.

Once the Fourth Army disaster team and other military units had departed, longer-term rehabilitation fell to volunteer organizations, particularly the Salvation Army, Volunteers of America, and the Red Cross.

Survivors still living today recall the efforts of the Salvation Army with particular respect. These and other charitable groups provided medical treatment, food and clothing, and shelter and helped anxious relatives with the perplexing task of locating individual survivors. Operating under its charter as a quasi-governmental disaster relief agency, a national Red Cross team of one hundred persons as well as five thousand local volunteers stayed on the scene for six months, expending a total of $1,361,000, $300,000 of which was for housing alone. While churches tended to focus upon the needs of their own congregations, an alliance of ministers met each morning to coordinate visits and assist those with physical or emotional problems.

Because the disaster commanded national media attention, the smoke had not even cleared before financial donations began arriving from all over the United States. Several committees were established to receive these, but foremost was the Texas City Relief Fund, chaired by Carl Nessler, a local banker and himself a former mayor. This was initiated by Mayor Trahan after a bank president in Houston informed him that the Busch family in St. Louis wished to contribute $50,000 for relief. Individuals and companies eventually donated $1,063,000 to this fund, which was paid out over several years' time. The fund was used to hire two social workers to counsel individuals, meet medical expenses, and pay the costs of career training for widows. A benefit performance by the New York Metropolitan Opera in Dallas, a Galveston recital by concert pianists Amparo and José Iturbi, and a show by Jack Benny also raised relief funds. Monsanto was quick to help its employees and widowed dependents. Two days after the *Grandcamp* blew up, W. T. "Andy" Anderson, who had narrowly escaped from the company's office building when the ship exploded, was busy making $1,000 cash payments to widows of Monsanto employees who had not been as lucky as he. Monsanto personnel at other facilities throughout the country raised a substantial sum to provide for the education of children who had a parent killed at the plant.

Insurance companies mounted an extensive program to handle policyholder claims for damaged residences. The National Board of Fire Underwriters immediately invoked its catastrophe plan, sending more than a hundred adjusters and supervisors to Texas City and Galveston. Approximately 4,000 claims were filed, 90 percent for losses on dwellings, and most were settled within three months. Oddly, most of the $3,955,000 paid out under fire insurance policies derived not from fire clauses but from an "extended coverage endorsement" for wind, explosion, and smoke damage taken out by policyholders with the hurricane threat in mind. Newspaper reports suggest that property insurance may have been difficult to obtain thereafter because companies had been obliged to pay out so much money.

Providentially for public morale and the economic future of Texas City, several companies which suffered heavily made immediate commitments to rebuild facilities or embark upon expansion programs. Even as the fires burned fiercely at Monsanto, Edgar Queeny, chairman of the board, came to Texas City and announced on the 18th that a new and expanded plant would be built on the site of the old one. Hourly employees were assured they would be kept on the payroll for this work, which ultimately required the better part of two years. Pan American Refining Corporation, which experienced only minor damage, purchased an additional seventy acres for expansion, and Republic Oil made plans to increase its refining capacity from 93,000 to 130,000 barrels daily. Carbide and Carbon Chemical Company, which lost nine storage tanks to fire, repaired its loading dock and pushed ahead with previous expansion plans. Insurance loomed large in commercial revival as well; it was estimated that the Oil Insurance Association, comprised of large stock fire insurance companies, had liabilities amounting to $50 million. Total outlay for industrial rehabilitation and development is estimated to have been about $100 million.

The Terminal Railway had a more difficult path to recovery. Although neither insurance companies nor the federal government would accept liability for $3 million in property damage, the company quickly cleared the South Slip of debris and began handling petroleum products by the end of April. A year passed before the rubble was entirely removed from the rest of the waterfront, and still another passed before the Army Corps of Engineers approved contracts to clear the north and Main Slips. Intermodal trade in railway boxcars resumed in July after Seatrain constructed a new crane at the end of the North Slip. But the net result was to hasten the shift of port business into petrochemicals and away from general cargo and agricultural products; the cotton compress and grain elevator were not rebuilt, and plans to resume break-bulk cargo were ultimately abandoned.

Flimflam

The Coast Guard Board of Investigation, which had been convened immediately after the disaster, finished its work in early May. A summary of its findings was released later that month, and the full report was issued in September. Testimony before the board as well as its own conclusions raised a number of possible causes for the ship explosions, but the board's

Wreckage of *Wilson B. Keene*, Main Slip. Moore Memorial Public Library, Texas City.

deliberations were seriously compromised by its failure to consider antecedent circumstances. The possibility of a more thorough investigation evaporated when the Secretary of the Treasury, head of the Coast Guard's parent organization, accepted the commandant's recommendation and convened an interagency committee to assess hazards associated with the transportation of the fertilizer; this effectively precluded consideration of the agency's responsibility for the safety of hazardous materials operations at ports, a fundamental reason for the disaster.

Under the circumstances, the board of inquiry was particularly disingenuous in recommending that the Justice Department prosecute the shipping agent for not notifying the French line of the fertilizer's characteristics and the French line itself for accepting damaged bags. In the very next paragraph of its report, the board noted that almost everyone was ignorant of Coast Guard regulations about the fertilizer. The board stated that this was so because military authorities had assumed control of the transportation of explosives only eight months after the regulations had been issued in April 1941.[4] What the board ignored was the Coast Guard's failure to fill the supervisory gap which developed after military control over munitions loading at ports was withdrawn in 1945. As discussed in Chapter 2, Admiral Shepheard, head of Merchant Marine Safety, while testifying under oath in *Dalehite v. United States*, admitted that hazardous cargo inspection was no more than a residual concern of the Coast Guard and that the Coast Guard captain of the port for Galveston had not known that ammonium nitrate fertilizer was being shipped through Texas City prior to the blasts. These awkward facts were never raised by the board and were neatly excluded from consideration by the interagency committee by confining its charge to technical matters.

The preliminary findings of the board, issued on 28 May, strongly suggest that, for his own part, Commandant Farley was content to apply a "Band-Aid" solution to the problems of handling hazardous materials in ports. He merely instructed shipping agents handling the fertilizer to properly prepare ships' holds in accordance with regulations and recommended that ships provide a fire watch during loading operations. Similar shortcomings existed in other ports. In the course of his investigation of the Texas City disaster on behalf of the National Board of Fire Underwriters, Matthew Braidech investigated fertilizer-loading operations at the Galveston wharves. He found lax safety practices there as well: "Every common fire safety rule was being violated with broken bags scattered about an unsprinklered warehouse."[5] Conditions were no different at the Port of Baltimore, a major outlet for fertilizer exports and the destination of railcars originally dispatched to Texas City. One other interesting fact was also never mentioned: in 1946, the Coast Guard had conducted tests which demonstrated that steam smothering was ineffective in putting out fires of combustible materials.[6] In sum, this agency's investigation and its subsequent actions strongly suggest that deficiencies in safety practices and supervision of hazardous materials operations were glossed over in order to deflect attention from its own culpability in the disaster.

Subsequent events were ironic in the extreme. On 5 July, the *Ocean Liberty* took on 3,300 tons of fertilizer at Baltimore, consigned to Brest, France. Twenty-three days later, while unloading in Brest harbor, this ship caught fire. At first, an attempt was made to smother the fire by closing or covering all portholes. This was unsuccessful, and a series of minor explosions followed. As the ship was towed away from the docks, it ran aground on a sandbar about half a mile from shore. An attempt to break open the hull by means of cannon fire failed. About 5:00 P.M., about three hours after it had run aground, black and red smoke shot up, indicating that petroleum barrels in an adjoining hold had caught fire. Then fire broke out in a second hold where fertilizer was located. At 5:25, the *Ocean Liberty* exploded, taking away the bow, killing twenty persons, and injuring about five hundred more. Waterfront facilities sustained $5 million worth of damage.[7]

At this point, some officials, including the New York City fire commissioner, concluded that the ship explosions at Texas City were not unique events and could happen elsewhere. The commissioner ordered two ships with fertilizer on board moored at the Brooklyn wharves to leave port immediately. The possibility that other ships carrying fertilizer would be embargoed at the foremost port in the nation finally impelled the Coast Guard to take meaningful action. It convened a conference on 6 August to discuss the dangers of handling fertilizer. Immediately thereafter, the commandant issued an order stating that ships were required to obtain a permit to load more than five hundred bags of fertilizer and then only at locations remote from concentrations of population or industry. In effect, the U.S. government had reverted to the wartime practice of confining fertilizer exports to facilities under military control, although several other terminals with strict supervisory procedures were added later. In September, the Interstate Commerce Commission amended rail and motor carrier regulations to require the yellow label for explosives on any product containing substantial amounts of nitrates. Two months later, this body also issued additional regulations and recommendations affecting land transportation and warehousing of the fertilizer.

Despite shoddy investigation and superficial attempts to affix blame, the Texas City disaster did generate considerable research and publicity about the explosive potential of ammonium nitrate fertilizer. In the latter part of the summer, Major General Philip Fleming, chairman of the President's Conference on Fire Prevention, convened a meeting of 120 officials and representatives of industry, transportation, and insurance to discuss the findings of the interagency committee. In turn, they established a committee of five persons to coordinate investigation into the properties and behavior of ammonium nitrate and its derivatives.[8] It was this group which uncovered the fact that the fertilizer had been involved in eleven serious explosions in the past.

Lawsuits and generous awards for damages are poor substitutes for thorough investigation of potentially hazardous situations and careful contingency planning for possible disasters. Nonetheless, the former do yield some insights into causes, including inadequate safety practices and poor supervision of operations. The Coast Guard's failure to enforce dangerous cargo regulations came to light in *Dalehite v. United States*. Consolidating 273 suits for damages related to the explosions filed under the Federal Tort Claims Act of 1946 on behalf of 8,485 persons, the claim by Elizabeth Dalehite and her son for the wrongful death of her husband and his father went to trial in 1949 before Judge T. M. Kennerly in the U.S. District Court, Southern Division of Texas. Millions of dollars were at issue, including substantial claims by insurance companies. Blaming almost everyone else, including the municipality, stevedore firms, longshoremen's unions, and shipping companies, the U.S. government denied having any responsibility for the deaths and injuries. Approximately twenty thousand pages of testimony and exhibits had been generated by the time Judge Kennerly rendered his verdict, just prior to the third anniversary of the explosions. He found for the plaintiffs, holding the United States at fault on some eighty specific points.[9] On appeal, this decision was overturned by the Fifth Circuit Court and confirmed on a four-to-three vote by the U.S. Supreme Court in 1953.[10] Both the Court of Appeals and the Supreme Court reached their decision on the basis of the meaning of culpability in the Federal Tort Claims Act of 1946. That is, the Supreme Court majority thought that the plaintiffs were not entitled to sue because the act confined liability to specific acts of negligence and not to "tortuous" conduct. In effect, the Court thought that Coast Guard responsibility for ensuring that dangerous cargoes were handled safely was meant by Congress to be an exclusive matter of governmental discretion.

The battle then shifted to the legislative arena. Clark Thompson, U.S. representative for Galveston, sought a special appropriation to compensate the victims. His bill, enacted by Congress in 1955, provided awards to individuals and not to underwriters and insurance companies. The Judge Advocate General of the Army was assigned the task of processing 1,755 claims and eventually made 1,394 awards totaling almost $17 million.

A New Perspective on Disasters

Although dilatory and of no use to victims of the disaster, the investigative efforts of the Coast Guard and other federal agencies did publicize the explosive qualities of ammonium nitrate fertilizer. In turn, these actions inspired more explicit regulations designed to increase the safety of transporting hazardous substances. Alerted to the drastic possibilities that might result from mistakes, officials probably took greater care in identifying and handling new chemical products. Even so, the evidence provided in *Dalehite v. United States* did not provide a complete explanation of the Texas City disaster. This is understandable in the sense that tort cases are designed to affix blame for the purpose of gaining reparation for wrongful death, personal injury, and property destruction. Inquiry in this case focused upon the immediate causes of the ship explosions in an attempt to establish who knew or should have known about the explosive properties of the fertilizer as well as upon safety practices at the docks. Fundamental and subtle questions about jurisdiction and responsibility for supervising hazardous materials operations received scant attention. Other matters affecting the severity of harm and damage at Texas City, particularly the absence of preparations for a serious industrial emergency, received no attention at all.

Devastation at the scene of the ship explosions; North Slip at left, Main Slip at right. National Archives.

Nonetheless, some persons did think about the implications that Texas City might have in terms of future events of this kind. A week after the disaster occurred, Texas governor Buford Jester was quoted in the *San Antonio Express* as hoping that the explosions would result in a "well-defined pattern for handling all aspects of future disasters."[11] In Texas City, some officials set out to effect improvements right away. One of these was Dr. Clarence Quinn, who had served as the mayor's medical coordinator during the crisis. He took the lead in formulating plans for an "emergency disaster unit" to facilitate acquisition of emergency supplies, provide expanded treatment, and designate transportation for casualties. Construction of the Galveston County Memorial Hospital near Texas City in 1949, financed by a large bond election, signified that attending physicians and the public had not forgotten their anguish over the fact that severely wounded victims of the *Grandcamp* explosion had to be sent to Galveston for adequate treatment. The Red Cross emerged from the disaster worried about assembling and delivering vast quantities of medical supplies on such short notice. Although the military had been able to collect and transport sufficient whole blood, plasma, and penicillin to avoid critical shortages during the first three days of the disaster, the difficulties involved prompted the National Red Cross to reevaluate its entire disaster relief program. Misunderstandings and resentment arising in the aftermath of the disaster likewise prompted the Red Cross to make a special effort to familiarize local authorities throughout the country with their responsibilities and resources in the event of a catastrophe. Undertakers were much the wiser as well. The Texas Funeral Directors and Embalmers Association submitted a plan, approved by the governor, that provided for a designated funeral director in the area of a disaster who would assume overall charge of handling the dead when local undertakers were unable to perform this task adequately.

Fire-fighting capabilities were improved as well. Refineries enhanced their capabilities by adding new staff and upgrading training and equipment. But their officials realized that individual company efforts would be insufficient if they were ever confronted by another catastrophe of similar magnitude. In 1948, H. M. Ross of Carbide and Carbon Chemical Company took the lead in setting up a fire-fighting cooperative called the

Industrial Mutual Aid System (IMAS). IMAS, which is still in operation, is coordinated by an executive committee of refinery and chemical company representatives as well as the municipal Fire and Police Departments. A manual identifies resources and sets forth procedures for providing assistance to any member whose resources are overwhelmed in an emergency and provides for periodic testing of plans through exercises. This organization is part of a larger arrangement for emergency preparedness conceived in 1947 that includes the Red Cross, industry, and government. IMAS became the lead agency for industrial emergencies, while civil defense handles military attack and the Red Cross covers natural disasters.

What had happened at Texas City also served as a catalyst for a similar and more extensive arrangement for natural as well as industrial disasters already under discussion in the Beaumont–Port Arthur–Orange area of southeast Texas. In February 1949, the Sabine-Neches Chiefs Association was established to support comprehensive mutual aid among the area's local and state law enforcement groups, municipal and company fire departments, local Red Cross and Salvation Army chapters, emergency program managers, and several federal agencies. This group and the Channel Industries Mutual Association, a large fire-fighting cooperative of industries along the Houston Ship Channel, remain viable entities today.

Although its primary reason was concern about Communist subversion, a few years after the Texas City disaster the Coast Guard took steps to improve the safety and security of munitions traffic through U.S. ports. At the direction of President Truman, it established the Port Safety Program in 1950. The following year it effectively reinstituted World War II port security standards for storage and transfer of "explosives and other dangerous articles," assigning Coast Guard captains of the port preeminent authority over all aspects of port safety and security within their jurisdictions.[12] These actions also affirmed regulations issued several months after the Texas City disaster that restricted loading of dangerous cargo to facilities remote from populated areas and only with a specific permit. Such facilities were required to provide trained guards; restrict welding, smoking, and movement of motor vehicles at the docks; and meet standards concerning electric wiring, fire extinguishing equipment, and storage of cargo.

While Texas City probably inspired some improvements in hazardous materials transportation and safety, the disaster may have had an even more profound impact—it may well have helped alter the way in which people thought about disaster preparedness. Although fragmentary, there is evidence suggesting that this was the beginning of a shift away from an essentially passive, reactive posture toward a more active, anticipatory mode. The evidence on this point is subtle and indirect, but clearly some individuals came to understand that when hazardous materials were at issue, the public interest required a community-wide effort to mitigate the worst effects of disaster and prepare response measures beforehand. There was a realization among some who had taken leading roles at Texas City that the almost exclusive reliance upon the Red Cross was misplaced in chemical disasters because of the complicated technical demands involved.

Associated Press correspondent Hal Boyle was among the first to articulate this theme. On 21 April, he filed a story that noted that while self-lessness prevailed among Texas City's citizens during the crisis, "widespread bitterness" was emerging over the lack of an organization capable of assisting communities disrupted far beyond the capabilities of charitable organizations and caring individuals.[13] This assessment was much the same as that of an investigator quoted earlier who noted that initial rescue and salvage efforts were much less effective because, without a contingency plan and prearranged mutual aid, they lacked necessary direction. Fire insurance expert Matthew Braidech grasped the larger implications of this weakness. "The Texas City Disaster also indicates the need for advanced preparedness to meet such critical hours," he stated. "Lack of planned organized relief also typifies many of our industrial communities. Authorities responsible for public safety should establish a major disaster plan gearing up all vital emergency services and arrange for mutual aid with nearby communal centers."[14] This theme was echoed by Colonel Homer Garrison, director of the Texas Department of Public Safety, who was also present at Texas City. In a long, rambling, but insightful speech about his experiences delivered to the International Association of Chiefs of Police, he asserted that response would have been much better if all activities had been integrated as well as police functions were. This, he noted, required more elaborate organization. "My suggestion is that the heads or top ranking officials of all participating agencies should be provided with some central place where they can all sit down together, all know what is going on, and thresh out . . . any problems of overlapping jurisdiction or duplication of effort. . . . This would be accomplished by having all emergency business clear the Control Center."[15] Reflecting on his experience in the disaster, Curtis Trahan told a reporter that "outside assistance should be coordinated through one local authority. This is the only way to avoid confusion."

Colonel Garrison grasped the essence of disaster management when he endorsed the concept of contingency planning for all kinds of emergencies. He reminded the audience that no one had expected Texas City to blow up, but it did. "With the threat of atomic war hanging over our heads," he continued, "I do not see how those of us who are responsible for public safety of the nation can do any less than to plan against every sort of disaster, whether from fire or flood or explosion or warfare."[16] His ideas were prescient in the sense that they envisioned the essentials of what was to become civil defense and later evolved into emergency management. He believed that towns and cities should have "at least the nucleus of a disaster organization with some safe and convenient spot arranged and equipped in advance to be used as a Control Center."[17] In turn, state governments should provide support and have a general plan for coordinating disaster activities, while the federal government should provide information and assistance.

Although the possibility of Soviet nuclear attack and passage of the Federal Civil Defense Act of 1950 were probably more important, Garrison's concerns about the difficulties at Texas City surely contributed to the passage of the Civil Protection Act of 1951 by the Texas Legislature.[18] This act put into place the fundamentals of a more consistent policy toward management of disasters. The governor was assigned responsibility for civil defense and disaster relief and authorized to create a Defense and Disaster Relief Council consisting of representatives of agencies and departments of state government and quasi-governmental agencies like the Red Cross. The council was to coordinate civil defense activity and support of mutual aid arrangements among local governments. Mobile Support Units were specified as providers of support services to areas hit by disasters, although counties and municipalities were encouraged to establish their own civil defense organizations. Since concern about nuclear attack was primarily at issue, legislation emphasized evacuation plans in the event of atomic attack and not preparedness for lesser emergencies, but it represented progress in that direction.

Again, with due allowance for the predominance of concern about a Soviet nuclear attack, Texas City may have also influenced government thinking at the national level. Supporting evidence appears in the report of a subcommittee of the National Research Council in March 1948. This report recommended that the U.S. Public Health Service create a permanent sanitary engineering unit to coordinate the efforts of numerous agencies at all governmental levels having relief responsibilities for civilian public health in case of war or major catastrophes.[19] Anticipating future developments, the report recommended that planning for sanitary engineering during major disasters should originate at the federal level and support detailed planning for local communities.

Deep personal loss lingered on because of injury or the death of loved ones, but with outside assistance and the initiative of its civic leaders, Texas City rapidly overcame much of the property devastation wrought by the explosions. Repair and reconstruction of housing were accomplished quickly, and most refugees were back in their homes within a few months. Retail business resumed normal operations within the same time frame. New housing was constructed to accommodate a steadily rising population. And, although break-bulk cargo operations never resumed at the port, restoration and expansion of petrochemical facilities spurred economic growth. While officials ignored or overlooked important causes of the disaster, the tragedy of Texas City, which had mesmerized the nation, nevertheless provided impetus for the first steps taken toward putting in place fundamental improvements in preparedness and response for industrial catastrophes, not only locally but at the state and national levels as well.

7

A Reckoning

Although the Dalehite case was replete with evidence of neglect and disinformation by responsible officials, these issues were never investigated. Indeed, as we have learned, the judgment of the district court against the United States, rendered on substantive grounds, was reversed by higher courts on an interpretation of governmental liability so restrictive that it would simply be unacceptable today. Even congressional hearings and an implicit admission of fault embodied in appropriation of funds as compensation to victims did not prompt further investigation.

How was it possible that everyone was ignorant of the catastrophic potential of the fertilizer, even though it contained ammonium nitrate? Most probably, blame belongs to officials of several federal agencies whose duties encompassed investigation and warning about hazardous materials. Foremost among these were the Interstate Commerce Commission (ICC), responsible for rail transportation safety, the Coast Guard, with Admiralty jurisdiction, and the Army Ordnance Bureau, which supervised the manufacturing process. Timely communication by the U.S. Bureau of Mines and the U.S. Department of Agriculture, both of which had some knowledge of the product, might have alerted others to the danger. In his "Findings of Fact and Conclusions of Law" for the Dalehite case, the district court judge wrote:

> All of Said Fertilizer stored on the *Grandcamp* and *High Flyer* was manufactured or caused to be manufactured by Defendant [the U.S. government], shipped by Defendant to Texas City, and caused or permitted by Defendant to be loaded onto such Steamships for shipment abroad. . . . All was done with full knowledge of Defendant that such fertilizer was an inherently dangerous explosive and fire hazard, and all without any warning to the public in Texas City or to persons handling same.[1]

Whether or not municipal or Terminal Railway officials were culpable, and even though a few engineers at local refineries and perhaps the municipal fire chief were uneasy about the fertilizer, these concerns could not overcome official lassitude.

Lax safety practices at the docks compounded the danger. If standards had been adequate, the fire on the *Grandcamp* might never have occurred, or one measure or another might well have interrupted the sequence of events which led to the ship's explosion. An absence of active federal supervision providing vital information, setting standards, and enforcing these by means of periodic inspections certainly contributed to local failure. As discussed in Chapter 2, the ICC did not require yellow warning labels on fertilizer bags. As far as the Coast Guard was concerned, safety measures affecting loading of hazardous materials on ships were "self-regulating and policing," and its captain of the port at Galveston was unaware that the ammonium nitrate fertilizer was being exported through Texas City before the explosions. Since the municipality did not assert jurisdiction over dock operations, safety was left by default to the Terminal Railway, whose short-term interest was to do relatively little.

Under these circumstances, the fire which started in the *Grandcamp* initiated a chain of events which caused the ship to blow up, inflicting destructive fury upon an unsuspecting town. The sequence of errors is reasonably clear. A stevedore's cigarette probably started the fire among the sacks in the *Grandcamp*'s hold because the no-smoking rule was not enforced. Then came the failure to use water to put out the fire and, instead, the master's critical mistake of sealing the hold and turning on the steam system. As the fire persisted and grew, all sorts of additional problems arose. No fire boat was present at the turning basin, and although the Terminal Railway lacked its own fire-fighting equipment, the Fire Department was not summoned for almost thirty minutes. Tugs were not even available to pull the *Grandcamp* away from the dock, and since turbine repairs were under way, the ship could not move under its own power. Finally, and most tragically, Dock O and Monsanto were not cleared of nonessential workers, and a crowd was allowed to gather at the head of the slip. The price of official ignorance about the fertilizer's explosive potential and the consequent surprise was the lives of several hundred innocent bystanders and serious injury to a similar number of people. Unfortunately, despite bravery and ingenuity on the part of those who responded, casualties and property damage were all the greater because Texas City was unprepared to cope with a major industrial disaster.

Preparedness and Response

Given the explosive power of ammonium nitrate, it could be argued that, prevention having failed, even the highest state of preparedness would not have averted much of the agony. The disintegration of the *Grandcamp* instantly created the direst sort of emergency. The entire area within a 1,500-foot radius of the blast epicenter became utter ruin, replete with masses of burning, twisted rubble, thick, heavy smoke, and noxious chemical fumes. Several hundred persons were felled, the dead and at least an equal number of seriously injured scattered about or trapped in building rubble. Many hundreds more with less serious injuries lay stunned or stumbled aimlessly about in severe shock. Beyond 1,500 feet, casualties were much less extensive, but damage was still heavy. As in a decapitated anthill, thousands of persons were impelled into urgent motion, everyone living in the flimsy houses south of Texas Avenue and east of Sixth Street compelled to flee homes collapsed by blast overpressure. Death, hospitalization, and flight during the 16th created the problem of locating two thousand missing persons which required two weeks to sort out. Property destruction in the form of crumbled ceilings and walls and glass from shattered windows and store fronts littering the streets extended at least a mile from the North Slip.

Whatever the immediate effects of the *Grandcamp*'s explosion, devastation was all the greater because no preparations were in place to quickly utilize outside assistance. Consequently, rescue efforts were disorganized during the early stages of response, and, with a few notable exceptions, little was available beyond what brave and resourceful individuals could accomplish. All four trucks of the Fire Department were destroyed, and half of its volunteer force were killed. Texas City had no hospital, and the three medical clinics sustained heavy structural damage. Utilities were knocked out of operation for varying periods of time. Because water mains were ruptured and fire trucks could not reach the dock area, nothing could be done about fires at the docks for several days. Many stubborn hydrocarbon fires at nearby tank farms could not be extinguished and had to be left to burn themselves out. Offices at the city hall, the logical center for coordinating response efforts, were heavily damaged, and the police radio was disabled for several hours. The local Red Cross chapter was itself victimized by the blast and, in the words of its leader, was "scattered all over town."

Because Texas City needed outside help to meet virtually all its needs, shortcomings in communications were a major impediment to effective response. All too typical of the first forty-eight hours after the explosion of the *Grandcamp* was the difficulty experienced by the Houston police captain who received Chief Ladish's telephone call. Lacking periodic updates about the situation, he had to mobilize help knowing only that a ship had blown up. As limited as communications with the outside world were, the problem originated within Texas City. Without a plan to allocate resources on a scale appropriate for such widespread carnage, operations were conducted in piecemeal, uncoordinated fashion. Rumors of impending explosions or toxic gas releases persisted throughout the 16th and 17th, inspiring periodic flights of citizens and disrupting search and rescue efforts at the docks. Not until 10:00 A.M. on the 17th—twenty-five hours after the *Grandcamp* disintegrated—was the mayor able to piece together a rudimentary staff, meet with army and Red Cross representatives, and lend relief efforts some coherence.

As noted above, there were important exceptions to overall chaos, for where organizational skills existed, response activities were performed quite well. This was particularly true of law enforcement and medical assistance. Officers from a number of surrounding jurisdictions and from nearby military bases converged on the town and, working under the direction of the police chief, quickly established access control and initiated patrols to prevent looting. Using hurricane plan guidelines, Galveston's physicians and nurses mobilized rapidly and began arriving in Texas City about an hour after the *Grandcamp* exploded. While a hospital at Texas City might have saved additional lives and afforded some injured persons better treatment, given that something like three thousand persons suddenly required medical assistance, the results are impressive. Other groups trained for emergencies also gave a good account of themselves, especially public utilities. But this was not true of fire suppression. During the 16th, refinery teams made limited progress suppressing fires at storage facilities, and little could be done at the waterfront until late on the 18th. The fact that the Houston Fire Department, largest in the area, arrived only after hearing alarming reports on commercial radio indicates that the confusion was never fully sorted out until most of the danger was over. More seriously, no one thought to check for additional hazards which the *Grandcamp*'s explosion might have created, one result being that the threat of the fertilizer on the *High Flyer* was overlooked until well past the time for effective remedial action. Indeed, nothing better illustrates the ramifications of the void in preparations than the virtual absence of contact between those operating at the turning basin and the municipal authorities on land.

It is not easy to ascertain who or what groups were chiefly responsible for the ship explosions and the ensuing devastation at Texas City. Consider the precursors, or background factors, discussed in Chapter 1—the close proximity of residences, petrochemical facilities, and waterfront operations; the social climate of indifference to the possibility of a catastrophe; and safety and preparedness for an industrial disaster. Although the Terminal Railway and the Mainland Company controlled industrial land use, proximity had developed over a long period of time. And while the public cannot be blamed for what happened, the awful results illustrate how prosperity which lulls people into complacency about risks can carry a heavy cost. Indifference to the safety of dock operations on the part of the municipality and the Terminal Railway Company is much more difficult to understand and more reprehensible. Large amounts of petrochemical products of known danger were congregated within half a mile of the waterfront. Toleration of lax safety practices at the docks by the municipality and union officials certainly amounted to simple negligence on the part of the municipality, the stevedoring company, and longshoremen's unions as well as culpable negligence on the part of the Terminal Railway as operators of the docks. Shipping agents, underwriters, and the master of the *Grandcamp* were culpable to the extent that they failed to observe pertinent safety standards.

But everyone's sense of danger and vulnerability depends upon the information he or she possesses. In this important respect, since ignorance of the dangers of ammonium nitrate fertilizer was fundamentally important, the ineptitude of several federal agencies was at issue, not the Terminal Railway, the unions, or the municipality. Given that ammonium nitrate was also a major element of munitions, it is inexplicable why the Army, Coast Guard, and ICC—and perhaps the Agriculture Department and Bureau of Mines as well—failed to consider its explosive potential and never warned those ports which handled the substance. Most especially, the Coast Guard took no interest and, by not filling the supervisory gap created when the Army curtailed its control over shipments of explosives through ports at the close of World War II, raised the opportunity for a hazardous materials transportation disaster.

The supervisory lapse on the part of the Coast Guard and its attempts to obscure what amounted to culpable negligence form the single most significant example of disinformation and disingenuousness connected with the disaster. Depositions by its officers in *Dalehite v. United States* as well as public statements by its officers serving on the board of investigation made three essential points: (1) within the meaning of *U.S. Code*, vol. 46, sec. 170 (12), in effect on 16 April 1947, the Coast Guard was obligated to inspect dangerous cargo only when violations were brought to its attention; (2) inspection of dangerous cargo operations was impossible anyway because of cutbacks in personnel after the war; (3) the Army's Bureau of Ordnance and the ICC failed to notify the Coast Guard of the dangerous potential of the fertilizer, and the latter failed to properly label the bags. The commandant contradicted the first assertion several months after the disaster when, on his own account, he ordered that fertilizer-loading operations be restricted to isolated sites and informed district commanders that they had general authority over the storage and stowage of explosives and dangerous substances on vessels. Only under oath and after meticulous and persistent questioning by counsel for the plaintiffs did any of its officers admit that the Coast Guard had not fulfilled its statutory duties with respect to supervising dangerous cargo. Concerning the second point, nowhere in any testimony or public statements did the Coast Guard claim that it had sought and been denied funding from Congress to improve inspection of loading operations of dangerous cargo at U.S. ports. Its assertion that the Army and the ICC were responsible for prevailing ignorance about the flammable and explosive qualities of the fertilizer amounts to more than disinformation—it had to be knowingly false. As noted in Chapter 2, not only did the Coast Guard designate ammonium nitrate as a dangerous substance in 1941, but it had access to all the research on the explosive potential of the substance developed during World War II.

The district court judge in *Dalehite v. United States* repeatedly labeled the actions of the Coast Guard and other federal agencies as "negligent" but not "culpably negligent." The latter designates willful or wanton behavior or inaction. It is ironic in the extreme that reversal of the district court decision by the Fifth Circuit Court of Appeals, based on a very narrow rendering of "tortuous conduct," had the effect of excusing Coast Guard inaction. It suggests that, at least in this instance, recourse to narrow interpretation of culpability by judges and their inability to place the actions or inactions of federal officials in proper context resulted in a failure of substantive justice.

As dreary as this litany is (and it could easily be extended), there is another side. Like any crisis, this one brought out the best in people as well as the worst. Thrusting aside surprise and shock at the sudden onslaught of horrible death, injury, and property devastation, many persons gathered up their courage and immediately tried to make things better, regardless of personal risk or sacrifice. Some officials performed far beyond their obligations, and those without obligations showed themselves willing to risk their lives or sacrifice time and money in order to ease the distress of victims. In retrospect, all sorts of images of compassion and courage emerge to assuage some of the pain and disappointment: of a young pupil helping his teacher, Rosa Curry, to her feet as her fifth-grade class fled from their school after the *Grandcamp* blew up; of hundreds of individual volunteers plunging into the maelstrom at the docks and working under dangerous circumstances for several days searching for the injured and dead; of the tireless efforts of Mayor Trahan, John Hill, Swede Sandberg, Chief Ladish, and others to organize relief programs; of the dogged determination of Lieutenant Sumrall, skipper of the buoy tender *Iris*; of local petrochemical companies and merchants providing supplies with no thought of compensation; of the quiet effectiveness of the Salvation Army; of the spontaneous generosity of Galvestonians; of rapid response by the Fourth Army and other military units from the surrounding area; of the long-term effort of the Red Cross to help citizens obtain housing; and of monetary contributions from people all over the nation, many of whom had never even heard of Texas City before 16 April.

Much was made in newspaper articles and popular journalism of Texas City's rapid economic recovery. In no small part, this is attributable to the fact that Edgar Queeny, president of Monsanto, quickly announced that the plant would be rebuilt. But there were other important achievements as well, such as the formation of the Industrial Mutual Aid System—the emergency cooperative of the municipality and industries—and better understanding in Texas City and elsewhere that advanced preparations were necessary to cope effectively with catastrophes. Particularly impressive is the fact that survivors—black, Hispanic, and white—remember that the experience of shared agony reduced barriers and increased understanding among the races and ethnic groups.

Nonetheless, given the frightful loss of life and property destruction, the good certainly did not outweigh the bad. Recall the quotation from Thucydides in the Preface that even if past mistakes are understood, they will be repeated in the future, human nature being what it is. Assessing the likelihood of another Texas City disaster is quite another matter and beyond the scope of this effort, yet it merits some comment. Fundamental lessons of this experience give us little reason for optimism. For one thing, given the destructive power of many products moving through commercial channels, if the awful consequences of explosion, fire, or toxic release are to be avoided, officials sharing responsibility for safety in U.S. ports must exercise unremitting vigilance in fulfilling their duties. The next hazardous materials disaster doesn't have to be different from Texas City—the advent of fertilizer shipments immediately transformed that port into a much more dangerous place, but nothing was changed, because no one bothered to investigate the fertilizer's explosive potential. Should the hazards of any new product be overlooked for very long, it becomes a matter of not *if* but *when* we will witness a catastrophe. Moreover, should public or private sector officials successfully exercise their natural inclination to hide culpability, then no one learns from their mistakes. The aftermath of the *Exxon Valdez* oil spill provides little cause for hope on this score. Although some sort of collective wisdom may emerge in dealing with industrial catastrophes, existing results suggest that disinformation, litigation, and threats of draconian penalties for repeat offenses are more likely products than useful lessons. In damage cases, candor usually takes a back seat to avoiding liability. Moreover, whatever the importance of human error as a cause of disasters, they are profoundly affected by the social situation and administrative arrangements as well.

Hazardous materials transportation is now comprehensively regulated, and safety technology is infinitely more sophisticated than in 1947. But the real question is whether these are adequate to cope with the presence of infinitely greater amounts of extremely hazardous chemicals and highly complex, sophisticated systems of production and transport. Most especially, can we rest assured that the present supervisory structure for the safety of ports and other aspects of maritime transportation does not have a gap equivalent to that created by Coast Guard inertia following World War II? Are we any less susceptible to the blandishments of prosperity than in 1947? Finally, there is the matter of how well ports and inland waterways can deal with major emergencies. Regulations now require that port authorities and companies have contingency plans for this purpose and that governments at all levels maintain some degree of emergency management capability. Nonetheless, recent oil spills demonstrate that fragmented jurisdiction—which helped obscure danger and hindered response in dealing with the fires on both the *Grandcamp* and the *High Flyer* —is still a problem. This is particularly worrisome at ports and navigable waterways, where the destructive effects of serious fires and explosions are likely to encompass both land and marine environments. At this juncture lurks the potential for another disaster on the order of Texas City.

Notes

Preface

1. Gary Walmsley and Aaron Schroeder, "Escalating a Quagmire: The Changing Dynamics of the Emergency Management Policy Subsystem," *Public Administration Review* 56 (1996):236.
2. Thucydides, *History of the Peloponnesian War.*

Acknowledgments

1. "Old Group Tensions Are Missing in Texas City, Reborn since Blast," *Washington Post*, 14 December 1947.

1. The Blasts

1. *Dalehite v. United States* (hereafter cited as *Dalehite v. U.S.*), U.S. Attorney, Precedent Case Files, Southern District of Texas, Houston Division, FBI Interviews, p. 471.
2. George Armistead, *Report to John G. Simmonds and Company, Inc., on the Ship Explosions at Texas City, Texas, on April 16 and 17, 1947, and Their Results*, p. 8.
3. G. M. Kintz, G. W. Jones, and Charles Carpenter, "Explosions of Ammonium Nitrate Fertilizer on Board the S.S. Grandcamp and S.S. High Flyer at Texas City, Tex., April 16, 17, 1947," p. 1.
4. Virginia Blocker and T. G. Blocker, Jr., "The Texas City Disaster: A Survey of 3,000 Casualties," *American Journal of Surgery* 78 (1949):759.
5. Ron Stone, *Disaster at Texas City*, p. 81.
6. Paul Shrivastava et al., "The Evolution of Crises: Crisis Precursors," *International Journal of Mass Emergencies and Disasters* 9 (1991):322.
7. P. Benham, "Texas City: Port of Industrial Opportunity," Ph.D. dissertation, University of Houston, 1987, p. 250.
8. F. Greenway, "The Chemical Industry," in Williams, ed., *A History of Technology: Vol. VI*, p. 534.
9. *New York Times*, 17 April 1947.
10. Leonard Logan, Lewis Killian, and Wyatt Marrs, *A Study of the Effect of Catastrophe on Social Disorganization*, p. 24.

11. Benham, "Texas City," p. 30.
12. National Fire Protection Association, "The Texas City Disaster: A Staff Report," *Quarterly* (July 1947):25.
13. "The Texas City Disaster," *National Fire News* 3 (May 1947):6.
14. F. D. Higbee, "How Ships Cause Port Destruction," *Log* (June 1947):24.
15. Ian Mitroff and Ralph Kilmann, *Corporate Tragedies.*
16. Logan, Killian, and Marrs, *Study*, p. 26.
17. Louise Comfort et al., "From Crisis to Community: The Pittsburgh Oil Spill," *Industrial Crisis Quarterly* 3 (1989):29.
18. Charles Perrow, *Normal Accidents*, p. 326.
19. Logan, Killian, and Marrs, *Study*, p. 25.
20. Shrivastava et al., "Evolution of Crises," p. 328.
21. National Fire Protection Association, "The Texas City Disaster," 32.
22. Logan, Killian, and Marrs, *Study*, p. 25.
23. Ibid.
24. Robert Kates, C. Hohenemser, and J. Kasperson, eds., *Perilous Progress*, p. 2.
25. Matthew Braidech, "The Texas City Fire and Explosion Disaster, Ammonium Nitrate Facts and Current Developments," *Proceedings, 74th Annual Conference of the International Association of Fire Chiefs* (hereafter cited as Braidech, *Proceedings*), p. 68.
26. U.S. Coast Guard, "The Texas City Disaster," *Proceedings of the Merchant Marine Council of the United States Coast Guard*, Washington, D.C., June 1947, p. 99.

2. The *Grandcamp*

1. Braidech, *The Texas City Disaster*, p. 16.
2. *Dalehite v. U.S.*, Defendant's Deposition, 20 January 1949, p. 37.
3. Braidech, *Proceedings*, p. 63.
4. U.S. Department of Agriculture, *Explosibility and Fire Hazards of Ammonium Nitrate Fertilizer*, circular no. 719, March 1945, p. 5.
5. U.S. House Committee on the Judiciary, *Hearings, Subcommittee no. 2 on S. 1077 and H.R. 4045*, 11 May, 8 June 1955, pp. 203–204.
6. Kintz, Jones, and Carpenter, "Explosions," p. 32.
7. Braidech, *The Texas City Disaster*, p. 6.
8. *Dalehite v. U.S.*, Transcript of Record 5:4746.
9. U.S. Senate, "Miscellaneous Report on Bill," Report 684, vol. 2, no. 11816.
10. Braidech, *The Texas City Disaster*, p. 16.
11. *Dalehite v. U.S.*, Consolidated Cases, Deposition of H. C. Shepheard, 20 September 1948, p. 35.
12. U.S. Coast Guard, "Coast Guard Final Findings at Texas City," p. 2.

13. *Dalehite v. U.S., Transcript of Record* 12:8807.

14. Braidech, *Proceedings*, p. 68.

15. *Dalehite v. U.S.,* Defendant's Deposition, 20 June 1949.

16. Braidech, *Proceedings*, p. 62.

17. *Dalehite v. U.S.,* FBI Interviews, p. 499.

18. *New Republic,* 28 April 1947.

19. *Dalehite v. U.S.,* FBI Interviews, p. 943.

20. Ibid., p. 332.

21. Ibid., p. 647.

22. Stone, *Disaster at Texas City,* p. 4.

23. A. A. Hoehling, *Disasters,* p. 119.

24. *Dalehite v. U.S.,* FBI Interviews, p. 843.

25. Kintz, Jones, and Carpenter, "Explosions," p. 11.

26. Ibid., p. 12.

27. *Dalehite v. U.S.,* Consolidated Cases, Defendant's Deposition, 20 January 1949, p. 17.

28. Braidech, *The Texas City Disaster,* p. 14.

29. U.S. Army, Ordnance Department, Picatinny Arsenal, "Explosibility of Ammonium Nitrate Fertilizer: Lecture by Wm. H. Rinkenback, 13 February 1948," p. 8.

30. *Texas City Sun,* 16 April 1991.

31. U.S. Coast Guard, "Record of Proceedings of Board of Investigation Inquiring into Losses by Fires and Explosions of the French Steamship *Grandcamp* and the U.S. Steamships *High Flyer* and *Wilson B. Keene* at Texas City, Texas, 16 and 17 April, 1947," *Proceedings,* Board of Investigation, p. 25. (Hereafter cited as USCG, Board of Investigation, *Proceedings.*)

32. Logan, Killian, and Marrs, *Study,* p. 22.

33. *Dalehite v. U.S., Transcript of Record* 13:11987.

34. D. Forrestal, *Faith, Hope, and $5,000,* p. 113.

35. James E. Bagg, "How Long Will You Remember?" *Texas Fireman* (May 1984):13.

36. *Dalehite v. U.S.,* FBI Interviews, p. 845.

37. Ivy S. Deckard, *In the Twinkling of an Eye,* p. 116.

3. Chaos and Courage

1. *Dalehite v. U.S.,* Transcript of Evidence 21:6257.

2. U.S. Army, Adjutant General's Office, "4th Army Report of Disaster Relief Activities by Military Personnel during the Texas City Disaster," prepared by Brig. Gen. J. R. Sheets, typescript, 8 May 1947, p. 2. (Hereafter cited as 4th Army, Disaster Relief Activities.)

3. U.S. Department of Interior, Bureau of Mines, "Letter Report: Texas City Disaster, Texas City, Texas" by E. J. Podgorski.

4. John H. Hill, "Plan for Disaster Control," *Chemical Engineering Progress* 43 (1947):11.

5. *Houston Chronicle*, 17 April 1947.

6. Ibid.

7. M. Garton, "Texas City," *Southern Funeral Director* 57 (July 1947):17.

8. *Houston Chronicle*, 16 April 1947.

9. Stone, *Disaster at Texas City*, p. 25.

10. Elizabeth Wheaton, *Texas City Remembers*, p. 16.

11. Ibid., p. 58.

12. Logan, Killian, and Marrs, *Study*, p. 41.

13. U.S. Coast Guard, "Radio Log of T-1218 and T-1896," typescript, 1947, p. 1.

14. USCG, Board of Investigation, *Proceedings*, p. 340.

15. Cliff R. Wattam, letter to Gene Tullich, 23 February 1992.

16. Patty Parish Hurt, "Recollections of Texas City," typescript, n.d., Galveston County Historical Commission, Rosenberg Library, Galveston, Texas.

17. 4th Army, Disaster Relief Activities, "Journal," Incl.4:4.

18. Logan, Killian, and Marrs, *Study*, p. 39.

19. Wheaton, *Texas City Remembers*, p. 58.

20. American National Red Cross, *The Red Cross at Texas City*, p. 10.

21. Bagg, "How Long Will You Remember?" p. 14.

22. Texas Adjutant General's Department, "Journal, Task Force Headquarters, Texas State Guard," 16 April 1947.

23. Roi B. Wooley, "Ammonium Nitrate Held Cause of Texas Disaster," *Fire Engineering* (May 1947):296.

24. H. Garrison, "The Texas City Disaster," *Illinois Policeman and Police Journal* (September–October 1947):32.

4. Struggling for Order

1. 4th Army, Disaster Relief Activities "Journal," Incl.4:5.

2. *Dalehite v. U.S.*, *Transcript of Record* 13:11953.

3. 4th Army, Disaster Relief Activities, "Journal," Incl.4:4.

4. U.S. Coast Guard, "Radio Log," p. 17.

5. George Tryon, "The Texas City Disaster," *Fireman* 1947, p. 2.

6. *Houston Chronicle*, 23 April 1947.

7. 4th Army, Disaster Relief Activities, "Journal," Incl.4:8.

8. 4th Army, Disaster Relief Activities, G-4 Report, 30 April 1947.

9. Wheaton, *Texas City Remembers*, p. 19.

10. USCG, Board of Investigation, *Proceedings*, p. 248.

11. *Dalehite v. U.S.*, FBI Interviews, p. 947.

12. Wattam, letter to Tullich.

13. *Dalehite v. U.S.*, FBI Interviews, p. 871.

14. Ibid., p. 872.

5. The *High Flyer*

1. *Dalehite v. U.S.*, Consolidated Cases, Defendant's Deposition, 10 January 1949.

2. *Dalehite v. U.S.*, FBI Interviews, p. 992.

3. USCG, Board of Investigation, *Proceedings*, p. 183.

4. *Dalehite v. U.S.*, Consolidated Cases, Defendant's Deposition, 20 June 1949, p. 38.

5. USCG, Board of Investigation, *Proceedings*, p. 251.

6. Ibid., p. 189.

7. Ibid., p. 252.

8. Ibid., p. 307.

9. *Dalehite v. U.S.*, FBI Interviews, p. 993.

10. U.S. Department of Interior, Bureau of Mines, "Letter Report," p. 2.

11. *Dalehite v. U.S.*, FBI Interviews, p. 707.

12. Stone, *Disaster at Texas City*, p. 65.

13. U.S. Department of Interior, Bureau of Mines, "Letter Report," p. 3.

14. Armistead, *Report*, p. 16.

15. Wheaton, *Texas City Remembers*, p. 23.

16. *New York Herald Tribune*, 18 April 1947.

17. Logan, Killian, and Marrs, *Study*, p. 50.

18. U.S. Department of Interior, Bureau of Mines, "Letter Report," p. 2.

19. Farrell Cross and Wilbur Cross, "When the World Blew up at Texas City," *Texas Parade* (September 1972):74.

20. M. Garton, "The Texas City Tragedy," *American Funeral Director* (July 1947):48.

21. Wheaton, *Texas City Remembers*, p. 21.

22. 4th Army, Disaster Relief Activities, "Journal," Incl.4: 6–7.

23. Texas Adjutant General's Department, "Journal, Task Force Headquarters, Texas State Guard," 16–21 April 1947, p. 79.

24. 4th Army, Disaster Relief Activities, "Journal," Incl.8:2.

25. *Houston Chronicle*, 20 April 1947.

26. *National Fire Protection Quarterly* (July 1947):35.

27. 4th Army, Disaster Relief Activities, "Journal," Incl.8:3.

28. *Lutheran Standard,* 21 June 1947.

29. *Houston Chronicle,* 21 April 1947.

6. Aftermath

1. American Red Cross, "Concise Report of the American Red Cross in the Texas City Disaster of April 16, 1947."

2. *Houston Chronicle,* 16 April 1967.

3. Armistead, *Report,* p. 41.

4. USCG, Board of Investigation, *Proceedings,* p. 549.

5. Braidech, *Proceedings,* p. 65.

6. Ibid., p. 109.

7. Ibid., p. 66.

8. Braidech, *The Texas City Disaster,* p. 16.

9. *Dalehite v. U.S.,* "Findings of Fact and Conclusions of Law Prepared and Filed by the Trial Judge under Rule 52 of the Federal Rules of Civil Procedure, Civil Action No. 787, 13 April, 1950," *Transcript of Record* 2:882–917.

10. 346 U.S. 15.

11. *San Antonio Express,* 24 April 1947.

12. *Federal Register,* 28 August 1951.

13. *Galveston Tribune,* 21 April 1947.

14. Braidech, *The Texas City Disaster,* p. 16.

15. Garrison, "The Texas City Disaster," p. 42.

16. Ibid.

17. Ibid.

18. State of Texas, Fifty-second Legislature, 1951, *Vernon's Civil Statutes,* art. 6889-4.

19. "Sanitary Engineers See Need of Catastrophic Organization," *Engineering News Record* 140 (1948):473.

7. A Reckoning

1. *Dalehite v. U.S., Transcript of Record* 2:2889.

Bibliography

Published Sources

American National Red Cross. *The Red Cross at Texas City*. St. Louis: American National Red Cross, Midwestern Area, May 1947.

Armistead, George. *Report to John G. Simmonds and Company, Inc., on the Ship Explosions at Texas City, Texas, on April 16 and 17, 1947, and Their Results*. N.p.

Bagg, James E. "How Long Will You Remember?" *Texas Fireman* (May 1984).

Blocker, Virginia, and T. G. Blocker, Jr. "The Texas City Disaster: A Survey of 3,000 Casualties." *American Journal of Surgery* 78 (1949):756–771.

Braidech, Matthew. *The Texas City Disaster: Facts and Lessons*. New York: National Board of Fire Underwriters, 1948. Moore Public Library, Texas City.

———. "The Texas City Fire and Explosion Disaster, Ammonium Nitrate Facts and Current Developments." *Proceedings, 74th Annual Conference of the International Association of Fire Chiefs*. New York, 1947. Moore Public Library, Texas City.

Comfort, Louise, J. Abrams, J. Camillus, and E. Ricci. "From Crisis to Community: The Pittsburgh Oil Spill." *Industrial Crisis Quarterly* 3 (1989):17–39.

Cox, George W. "Observations Relating to the Texas City Disaster." *Industrial Medicine* 16 (1947).

Cross, Farrell, and Wilbur Cross. "When the World Blew up at Texas City." *Texas Parade* (September 1972):70–74.

Deckard, Ivy S. *In the Twinkling of an Eye*. New York: Vantage Press, 1962.

"Doctors Tell Story." *Medical Economics* 24 (June 1947):70–75.

Forrestal, D. *Faith, Hope and $5,000: The Story of Monsanto*. New York: Simon and Schuster, 1977.

Garrison, H. "The Texas City Disaster." *Illinois Policeman and Police Journal* (September–October 1947):18–42.

Garton, M. "Texas City." *Southern Funeral Director* 57 (July 1947):16–19.

———. "The Texas City Tragedy." *American Funeral Director* (July 1947):27, 48, 68–70.

Greenway, F. "The Chemical Industry." In T. Williams, ed., *A History of Technology, Vol. VI.* Oxford: Clarendon Press, 1978.

Higbee, F. D. "How Ships Cause Port Destruction." *Log* (June 1947):17–20.

Hill, John H. "Plan for Disaster Control." *Chemical Engineering Progress* 43 (1947):11–15.

Hoehling, A. A. *Disasters.* New York: Hawthorne Books, 1975.

Kates, Robert, C. Hohenemser, and J. Kasperson, eds. *Perilous Progress: Managing the Hazards of Technology.* Boulder, Colo.: Westview Special Studies, 1985.

Kintz, G. M., G. W. Jones, and Charles Carpenter. "Explosions of Ammonium Nitrate Fertilizer on Board the S.S. Grandcamp and S.S. High Flyer at Texas City, Tex., April 16, 17, 1947." *Report of Investigations, 4245.* Washington, D.C.: U.S. Department of Interior, 1948.

Logan, Leonard, Lewis Killian, and Wyatt Marrs. *A Study of the Effect of Catastrophe on Social Disorganization.* Chevy Chase, Md.: Operations Research Office, Johns Hopkins University, 1951.

Mitroff, Ian, and Ralph Kilmann. *Corporate Tragedies: Product Tampering, Sabotage and Other Catastrophes.* New York: Praeger Special Studies, 1984.

National Fire Protection Association. "Recent Port and Ship Fires and Explosions." *Quarterly* (October 1947):109–121.

———. "The Texas City Disaster: A Staff Report." *Quarterly* (July 1947): 27–57.

"Old Group Tensions Are Missing in Texas City, Reborn since Blast." *Washington Post,* 14 December 1947.

Orr, D. "Catastrophe and Social Order." *Human Ecology* 7 (1979):41–52.

Perrow, Charles. *Normal Accidents: Living with High-Risk Technologies.* New York: Basic Books, 1984.

Reed, Roy G. "How a Health Department Functions in a Disaster." *Texas Journal of Public Health* 24 (1947):5, 9–12.

"Sanitary Engineers See Need of Catastrophic Organization." *Engineering News Record* 140 (1948):473–475.

Shrivastava, Paul, Ian Mitroff, Dan Miller, and A. Miglanti. "The Evolution of Crises: Crisis Precursors." *International Journal of Mass Emergencies and Disasters* 9 (1991):321–327.

Stone, Ron. *Disaster at Texas City.* Fredericksburg, Tex.: Shearer Publishing, 1987.

Texas Legislature. 1951. *Civil Protection Act.* H.B. no. 787. *Vernon's Annual Civil Statutes,* chap. 311, art. 6889.

Thucydides. *History of the Peloponnesian War: A Companion to the English Translation.* Trans. Rex Warner. Bristol: Bristol Classic Press, 1985.

Tryon, George. "The Texas City Disaster." *National Fire News* 359 (1947): 2–3, 6.

———. "Texas City Disaster." *Fireman* 1947 (reprint).

U.S. Coast Guard. "Record of Proceedings of Investigation Inquiring into Losses by Fires and Explosions of the French Steamship *Grandcamp* and the U.S. Steamships *High Flyer* and *Wilson B. Keene* at Texas City, Texas, 16 and 17 April, 1947." *Proceedings,* Board of Investigation. Washington, D.C., 1947, pp. 98–99.

———. "The Texas City Disaster." *Proceedings of the Merchant Marine Council of the United States Coast Guard.* Washington, D.C., June 1947, pp. 98–99.

U.S. Department of Agriculture. *Explosibility and Fire Hazards of Ammonium Nitrate Fertilizer.* Circular no. 719, March 1945.

U.S. House Committee on the Judiciary. *Hearings, Subcommittee no. 2 on S. 1077 and H.R. 4045,* 11 May, 8 June 1955.

———. "Statement of W. H. Sandberg, Employee of Texas City Terminal Railway Co." *Hearings on H.R. 296,* 16, 17, 18 November 1953, 76–88.

———. "Supplementary Statement of the Department of Justice on S. 1077 and H.R. 4045, Texas City Claims Bills." 84th Cong., 11 May, 8 June 1955.

U.S. Senate. *Hearings before a Subcommittee of the Committee on the Judiciary, United States Senate, 84th Congress on S. 1077.* 17 May, 7 June 1955.

———. "Miscellaneous Report on Bill." Report 684, vol. 2, no. 11816.

"Utilities Quick Action Restores Electric Service at Texas City." *Electrical World*, 26 April 1947.

Walker, D. "A Texas City Local Agent's View of the Historic Disaster." *Local Agent* (June 1947):10–11, 41–42.

Walmsley, Gary, and Aaron Schroeder. "Escalating a Quagmire: The Changing Dynamics of the Emergency Management Policy Subsystem." *Public Administration Review* 56 (1996):235–244.

Wheaton, Elizabeth. *Texas City Remembers.* San Antonio: Naylor, 1948.

Wooley, Roi B. "Ammonium Nitrate Held Cause of Texas Disaster." *Fire Engineering* (May 1947):294–301, 314.

Unpublished Sources

American Red Cross. "Concise Report of the American Red Cross in the Texas City Disaster of April 16, 1947." Rosenberg Library, Galveston, Texas.

Archival Records, the Texas City Disaster. Moore Public Library, Texas City.

Benham, P. "Texas City: Port of Industrial Opportunity." Ph.D. dissertation, University of Houston, 1987.

Braidech, Matthew. "Recent Developments in Ammonium Nitrate Fertilizer." Paper presented to 52nd Annual Meeting of National Fire Protection Association, Washington, D.C., 12 May 1948.

Browne, H. F. "Letter Report: Texas City Disaster, Texas City, Texas." Docket of Consolidated Texas City Disaster Cases, technical data and reports, May 1947. RG 118, National Archives, Southwest Region, Fort Worth, Texas.

Carbide and Carbon Chemical Company (Union Carbide). "Relief Organization: Preliminary Report, April 22, 1947." In author's possession.

Cruickshank, J. W. "Texas City Disaster." RG 118, National Archives, Southwest Region, Fort Worth, Texas.

Fire Prevention and Engineering Bureau of Texas and the National Board of Fire Underwriters. "Texas City, Texas Disaster, April 16, 17, 1947." Moore Public Library, Texas City.

Hurt, Patty Parish. "Recollections of Texas City." Typescript, n.d. Galveston County Historical Commission, Rosenberg Library, Galveston, Texas.

KPRC News Department. "Texas City Diary Apr. 16 thru Apr. 18, 1947." Archival Records, the Texas City Disaster. Moore Public Library, Texas City.

Stephen, W. "The Texas City Disaster." 1947. Rosenberg Library, Galveston, Texas.

Texas Adjutant General's Department. "Journal, Task Force Headquarters, Texas State Guard." 16–21 April 1947. National Guard Museum, Austin, Texas.

U.S. Army, Adjutant General's Office. "4th Army Report of Disaster Relief Activities by Military Personnel during the Texas City Disaster." Prepared by Brig. Gen. J. R. Sheets. Typescript, 8 May 1947. RG 370.1, National Archives, Washington, D.C.

U.S. Army, Corps of Engineers, Galveston District. "Brief Report of Texas City Disaster." Typescript, 24 April 1947. RG 407, National Archives, Washington, D.C.

U.S. Army, Ordnance Department, Picatinny Arsenal. "Explosibility of Ammonium Nitrate Fertilizer: Lecture by Wm. H. Rinkenback, 13 February 1948." RG 118, National Archives, Southwest Region, Fort Worth, Texas.

U.S. Attorney, Precedent Case Files, Southern District of Texas, Houston Division. "Cases Relating to the Texas City Disaster. Transcript of Evidence, Exhibits, Depositions, Pleadings, FBI Reports, Technical Data and Reports." RG 118, National Archives, Southwest Region, Fort Worth, Texas.

U.S. Coast Guard. "Coast Guard Final Findings at Texas City." 1947. RG 118, National Archives, Southwest Region, Fort Worth, Texas.

————. "Preliminary Findings at Texas City." 1947. RG 118, National Archives, Southwest Region, Fort Worth, Texas.

————. "Radio Log of T-1218 and T-1896." Typescript, 1947. Historical Division, Coast Guard Headquarters, Washington, D.C.

U.S. Department of Interior, Bureau of Mines. "Letter Report: Texas City Disaster, Texas City, Texas" by E. J. Podgorski. Docket of Consolidated Texas City Disaster Cases: Technical Data and Reports. RG 118, National Archives, Southwest Region, Fort Worth, Texas.

U.S. Department of the Treasury. "Press Service No. S-446." 30 August 1947. Archival Records, the Texas City Disaster. Moore Public Library, Texas City.

U.S. Navy, Eighth Naval District. "Navy Assistance at Texas City." Letter to Mildred Stevenson, 17 September 1947. Rosenberg Library, Galveston, Texas.

U.S. President, President's Conference on Fire Prevention, Conference on Fire Hazards Involved in the Manufacture, Transportation, Storage and the Use of Ammonium Nitrate. "Proceedings." 8 September 1947. Washington, D.C. Moore Public Library, Texas City.

Wattam, Cliff R. Letter to Gene Tullich, 23 February 1992. In author's possession.

Index